THE CALIFORNIA CAULDRON

THE CALIFORNIA CAULDRON

Immigration and the Fortunes of Local Communities

William A. V. Clark

University of California, Los Angeles

THE GUILFORD PRESS

New York   London

© 1998 The Guilford Press
A Division of Guilford Publications, Inc.
72 Spring Street, New York, NY 10012
http://www.guilford.com

Printed in the United States of America

This book is printed on acid-free paper.

Last digit is print number: 9 8 7 6 5 4 3 2 1

Library of Congress Cataloging-in-Publication Data

Clark, W. A. V. (William A. V.)
    The California Cauldron : immigration and the fortunes of local
communities / William A. V. Clark.
        p.   cm.
    Includes bibliographical references and index.
    ISBN 1-57230-403-0 (hardcover)
    1. California—Ethnic relations.   2. Immigrants—California—
Social conditions.   3. Immigrants—California—Cultural
assimilation.   4. Community life—California.   5. California—
Population.   6. California—Emigration and immigration.   7. Human
geography—California.   I. Title.
F870.A1C53   1998
305.8'009794—dc21   98-38943
                                                                      CIP

CONTENTS

LIST OF FIGURES AND TABLES

Figures

Tables

PREFACE

In the first half of the 20th century, California was the "promised land," and the state grew and prospered with expanding economic opportunities and burgeoning migration. The newcomers came mostly from the Midwest; they were largely white and Protestant, and although Hispanics and Asians lived in California, in the years after World War II it was still an "anglo" state.

Beginning in 1965, however, changes in the immigration law altered the influx from abroad and transformed the nation, especially California. Once a destination for young migrants from within the United States, California changed to become a worldwide magnet. The California of the last two decades of the 20th century differs markedly from the California of image and perception in the 1960s. The state is now the promised land for a new wave of immigrants and is populated by a very different racial and ethnic mix. How will these new waves of immigrants mesh with the existing population?

In this book I discuss how the blending will occur across the geography of California. What are the geographic outcomes of this fundamental transformation of the state's population? What is happening now, and what can and might happen, in California's cities, towns, and neighborhoods as the new migrants take up their lives in a new and different society? I explore the constraints, problems, and implications of the intersection of 5 million newcomers in the past 15 years with the 25 million residents already in the state.

Although the debate about immigration and the perceptions that fuel that debate unfold on the national stage, the consequences of immigration itself are focused sharply on certain places. The process of immigration operates like a giant parabolic mirror, collecting newcomers from around the world and concentrating them in particular communities, many within California. Although this book examines the trajectories of immigrant successes and failures in California, the major emphasis is on the way in which influxes of newcomers from around the world locate themselves in particular communities and neighborhoods in California. What have the local outcomes been? Who gains? Who bears the burdens? The question is not simply whether immigration is problematic. The question and the answers depend on who the immigrants are, where they come from, and where they settle. Immigration benefits some immigrants and some communities; it also imposes burdens unequally.

The picture I paint in the following chapters is necessarily complex. It reflects the varying experiences of both the immigrants and the communities where they settle, as the newcomers struggle with transforming themselves into citizens of a new society. It is clear, however, that the current laissez-faire immigration policy needs a major revision. At the very least, this study shows that a rational immigration policy would protect the less able citizens who are already in this country, and would establish clear rules for admitting additional immigrants who would increase the country's overall human capital. At the same time, we ignore at our peril the very large number of new citizens who are at the very bottom of the economic ladder and who have little opportunity for experiencing upward mobility. Thus we have a dual need: to limit immigration and to educate those who are already here. At the moment, neither is happening.

The book relies heavily on U.S. Bureau of the Census and other statistical information to make arguments about the immigration process and its outcomes. Necessarily, this approach pays less attention to the individual stories of migrants coming to a new land. However, this "quantitative" approach allows me to provide a wide range of findings on just how immigration is intersecting with local communities. By measuring the changes, I am able to show that much of the recent growth in California is now being created by immigration, and not by natural increase from the native-born population; and that there has been a major drop in the proportion of births to United States–born mothers. Now, in the last few years of the 20th century, California is replicating the United States as a whole at the beginning of the century. A quarter of the population are foreign-born, and nearly one-third of the population speaks a language other than English at home. This fundamental change in the composition of the California population has implications that I explore in detail in the substantive chapters. The statistics on wages and incomes show that even though immigrants are losing ground against white native-born workers, they are often earning about the same in real dollars as their counterparts were earning when they came to the United States decades ago. It is not hard to understand the continuing attraction of California for migrants.

There is an overriding feeling that the "American Dream" of home-ownership and upward educational mobility is alive and well in the "New California." Homeownership levels are very high for many immigrants, and immigrants seem to be pushing up, and, when they can, moving out to suburban cities and communities. At the same time their story is not always rosy. Mexican and El Salvadoran immigrants are losing ground; their more limited education and low skills make it hard for them to take a central role in an increasingly technological society. It is the very large numbers of these immigrants that are at the center of the problem of how to create a new and integrated society.

This book is the outcome of a two-decade-long interest in human mobility and migration at local and regional scales, and of the realization that transformations under way in the nations metropolitan areas are being shaped—directly and indirectly—by immigration. A continuing flow of new immigrants from Mexico and Asia, the ongoing high levels of fertility, and the new ethnic citizen children assure a situation in which foreign ethnic identities will challenge our conceptions of diversity and assimilation.

*   *   *

Some of the work in this book has been published previously in *Regional Studies*, in the *International Journal of Population Geography*, and in conference proceedings. I am grateful for permission to reprint figures, diagrams, and tables. I also wish to acknowledge the assistance of my able, longtime computer collaborator, Jeffrey Garfinkle, who has been involved in my migration and housing research for the last decade. The diagrams and maps are the creative contribution of Chase Langford, the cartographer in the Department of Geography at UCLA. I wish to thank the John Simon Guggenheim Foundation for the generous support that enabled me to begin this project. A number of friends and colleagues, notably James Allen, Peter Morrison, and Roger Waldinger, were kind enough to read large parts of the manuscript and to offer detailed and very helpful advice. The book owes much to their patience and insight and to that of my wife, Irene.

CHAPTER 1

INTRODUCTION:

MIGRATION IN A CHANGING WORLD

The United States proclaims itself a nation of immigrants, a society formed and created by successive waves of people from other lands, who arrived in especially large numbers in the first two decades of this century. This claim, however, pays little attention to U.S. government efforts, beginning in the 1920s, to severely restrict immigration or to the often vociferous contemporary efforts to totally halt immigration. For almost half a century, after the restrictions of the 1920s, the influx of newcomers was closely controlled and relatively restricted. Since 1965, however, immigration has undergone a sea change, and nowhere is it more visible than in the supermarkets, taxicabs, hotels, and schools of postindustrial California.

The state's population is now more than one-quarter foreign born, and the principal issue confronting California today is whether this polyglot society can work. Nor does this question apply only to California; what happens in California tomorrow is likely to happen in the nation as a whole the day after tomorrow. Those who see recent changes in a favorable light emphasize how the California economy is reinventing itself yet again, enjoying solid growth in a wide variety of service and professional specialties. Because immigrants have provided important low-cost labor to fuel this growth, the proponents of immigration argue that such a picture augers well for the nation. Another view of the same picture, however, suggests increasing social inequality, social tension, separation, and balkanization as the consequences of the rapidly changing population composition of the state.

This foreign influx has drawn widespread attention and has spurred considerable controversy. The 1970s and 1980s have been described as the decades of "mass migration" and the late 20th century as the "age of migration" (Castles and Miller, 1993). As immigration has returned to record historic heights, the debates over immigration policy have intensified at both the national and local levels. Moreover, the changes in the ethnic and racial composition of the migration streams have raised issues about the social and political implications of contrasting cultures joining the American mainstream. The increasing size of the migrant flows and the difference in their demographic composition have interacted to prompt a broad, sometimes contentious, debate on the issue of immigration. The debate is most intense and most vociferous in California, where the size and composition of the flows, and the problem of illegal immigrants, have been the topics of academic reports, political debates, and media discourse.

Public attention is now focused firmly on the size, nature, and implications of continuing large-scale migration into the United States.[1] The debates over immigration reflect a growing concern about the consequences—both real and perceived—for California and the nation (Martin and Midgely, 1994). Is the migration a temporary adjustment, or will the new immigrants fundamentally change U.S. society? Will U.S. institutions and organizations be able to adapt and change to re-create a new *United* States, or will our society "balkanize" along ethnic lines, socially and geographically? Will Soviet Armenians, Asian Indians, Somalis, and other recently arrived groups create their own versions of "Chinatown"? Already, some Californian cities contain intense concentrations of Vietnamese, Cambodians, and Koreans. These questions are important, and although there is a large literature on immigration, notably the recent study by the National Research Council (1997), *The New Americans*, much of it does not address the geography of immigration—of what is actually happening in states and communities and neighborhoods. Yet the issues concerning the nature of changes in our society are most palpable and most intense at the local level. The changes in neighborhood school systems, in state electoral districts, and in city and county politics are at the heart of the day-to-day concerns of the citizens in communities in California, and in other states with burgeoning immigrant populations.

Much of the national debate concerns the costs and benefits of immigration, but in the end these debates may be unresolvable. Evaluating costs and contributions has an important temporal dimension. Costs may accrue here and now in a particular place or region, but any economic gains stretch out over a much longer term as the new immigrants enter the workforce, pay taxes, and contribute to the country's productivity. Moreover, the gains may register in some other place, or in the nation as a whole (e.g., through federal taxes paid by workers). Thus I emphasize the way in which immigrants are affecting particular locations and focus on the changes that are occurring among immigrants in those locations. The heart of this book is about the geography of immigrants in California.

MIGRATION AND MIGRANTS

International migration has three dimensions: the flow of migrants to a particular destination, the immigrant population already living at that destination, and the flow of those who return to their places of origin. Not all migrants stay in their new country, and the number who leave to return to their original homeland has varied by time and place. The fact that the return flow of migrants has declined dramatically may be an important key to understanding the recent buildup of migrants in California. Nor does international migration occur in a vacuum: it is tied closely to the economic and social situation in both the country of origin and the country of destination.

Demographers and other social scientists have developed a set of explanatory conceptualizations to organize our understanding of migration in terms of the economic and social contexts of origins and destinations that motivate and channel migrants' directions. Earlier theories of international migration emphasized differences in economic opportunities between place of origin and destinations: jobs offering higher wages and greater opportunities for economic advancement were the key to understanding flows of immigrant labor from one country to another. According to this model, areas of high wages and high labor demand are the magnets that "pull" newcomers, and areas of low wages and low demand are the "pushes" that motivate departure. In this neoclassical economic interpretation of migration, the flows of immigrants from low-wage to high-wage areas reflect, in aggregate, the choices of individual migrants, each of whom is striving to improve his or her individual position in society. This explanation of migration emphasizes the role of improving human capital by migrating to obtain greater economic opportunities. According to this view, the choices people make about where to live and work reflect the average differences between regions and nations.

A substantial research literature focuses on the role of economic incentives and opportunities in promoting international migration. The central idea is that individuals move to improve their human capital, they invest in themselves, so to speak. The 20-year-old single man from a village in Mexico who moves to a low-wage job in Los Angeles is moving in the expectation of being better off in the long run. The hope of an increased income stream over his lifetime becomes a primary motivating force in his decision to move (DaVanzo, 1981; Mincer, 1978; Sjaastad, 1961). This idea of improving human capital provides important insights, including the notions that benefits from migration do not occur instantaneously; that individuals can move even if no immediate returns are forthcoming; and that the *expectation*, not the certainty, of being better off is a critical dimension of the process. In the human capital conceptualization, migration is an investment that incurs immediate costs, which are balanced against expected future returns. Migrants make a personal investment in their ability to be more productive, and thereby generate a greater lifetime income, in their new location.

Economic opportunities loom large in motivating long-distance migration. Although migrants know that the streets in California are not paved with gold, it is still true that the difference between extremely low or nonexistent wages in many countries of origin and the low wages of the postindustrial cities in California is an important incentive. At the same time, while the neoclassical economic theories suggest that migration is impelled by economic "pull" and "push," and that in time migratory flows tend to lessen economic differences between regions and nations, we know that migration occurs even when the opportunities are limited and when jobs may not be available at all.

Other social scientists have emphasized that economic models are incomplete explanations for the patterns of migration. They point out that some poor nations have not generated significant migratory flows to the United States, and that it is not always the poorest and most dependent people who emigrate (Boyd, 1989; Portes and Rumbaut, 1996). The migrants are often those with middle levels of social status. None of these observations negate the human capital explanation of migration, but they do suggest that it is incomplete.

To explain why the flows of migrants continue when jobs in the destination country are low-paying or do not exist, students of international migration have invoked the idea of networks and contacts as additional reasons. Immigrant communities, once they are established, offer an entry point into the new society and an initial set of contacts for new migrants. Social and personal links become important in fostering and enabling continuing migration even when no job is immediately available for a new arrival. Thus flows that formerly may have been stimulated by economic differences will acquire a momentum of their own, and will continue even when economic conditions would not seem to encourage such flows. In Orange County, for example, a Mexican village in effect has been re-created in Santa Ana. Over a 30-year period, the initial migrants from the village of Granjenal have re-created that community in a new geographic setting. The increasing stock of immigrants in particular destinations perpetuates flows by expanding the network of links between the migrants and the potential migrants still living at the place of origin (Waldorf, 1996). We can expect to see more immigrants in those communities in California where there are already substantial numbers of earlier arrivals from other countries.

The role of networks typically amplifies and gives geographic focus to the more impersonal economic forces that motivate immigration. Thus the human capital model of migration is not inconsistent with a model that emphasizes the importance of previous migration in the continuing flows (Hugo, 1981; Kritz and Zlotnik, 1992; Massey, Donato, and Liang, 1990). We know that migration is "channelized," that is, migrants from particular locations are likely to settle in particular destinations. The settlement of the Hmong hill people from Vietnam in Fresno, California, and later in St. Paul, Minnesota, exemplify the effects of chain migration. Pioneer migrants find their way to destinations where they thrive; word goes back to other would-be migrants; and news of the success of earlier migrants generates additional flows to these destinations. As a result, particular migrant groups are concentrated in particular communities (Miyares, 1997).

Both economic processes and social networks influence the flows of migrants to the United States. We have only to drive along the streets serving the construction trades and observe the clusters of "day laborers" to see the realities of economic opportunity. Such opportunities continue to draw migrants, especially from nearby Mexico; there the wages are much lower and the opportunities are fewer, especially in the rural communities. Opportunities, however, extend beyond labor markets to government support programs that enable families to bring their aging parents to the United States. Again, the role of networks and social links is critical. Changes in U.S. immigration law, with the dual effect of favoring skilled migrants and family reunification, in combination with large existing concentrations of immigrants, have significantly increased the flows of migrants from Mexico and certain Southeast Asian nations into the United States (Martin and Midgely, 1994).

The above discussion of economic incentives and social networks emphasizes migrants' individual behavior, the nature of their decision making, their assessment of relative costs and benefits, and their connection to other groups of migrants. Yet, although the idea of individual migrants making deliberate choices is significant, it is also important to recognize the role that national policies play in creating, sustaining, and influencing migration. Nation-states play major roles in shaping and controlling migrants' movements. Immigration was a major part of nation building in Australia, Canada, and New Zealand. Nation-states also have played a major role in labor recruitment as in the "guest worker" programs in Germany after World War II and the "bracero" program in the United States which recruited agricultural workers from Mexico. As the world's political economy has changed, so has the nation's political involvement in alternately stimulating and restricting migration. Thus, although individual migrants make personal decisions regarding the opportunities and costs they perceive, and although they use the networks and contacts established by earlier migrant groups, their decisions are set in a larger context of changing national policies that may facilitate or impede their intended actions.

The migratory process is complex and multifaceted, and its complexity has made it difficult to control and influence. Migrants are not a unified group; and they come at different times and for different reasons. Understanding some of the nature of migrants and their historical background provides a context for examining the specific changes in California.

TYPES OF MIGRANTS

Historically, persons who move to work as unskilled laborers in agriculture, construction, services, or even industry have accounted for the bulk of new immigrants, especially in California. These labor migrants fit most closely with our conception, whereby migration flows in response to differences in economic disparities. The lowest paying jobs in the United States pay at least four times as much as even reasonably paying jobs in Mexico. These labor migrants can enter the United States legally with a tourist visa and simply "overstay," or, more often, enter without any documentation. Illegal border crossing has become frequent and large-scale in the last two decades; those who cross the border illegally or enter without inspection (EWIs) are now a very large proportion of all immigrants, which the 1986 Immigration Reform and Control Act (IRCA) was designed to discourage. Many labor migrants also enter as part of the family reunification preference structures, in which legal foreign-born residents may be entitled to bring in family members from abroad.

The labor flows are clearly related to demand. Although family reunification is an important part of the migration process, many migrants would not come if there were not a demand for their labor. The demand appears to be strong and sustained. Employers laud the immigrants' industriousness and reliability and often argue that without inexpensive labor they would have to move jobs abroad, or, in the case of agriculture, let their crops go unpicked. At the same time, many of these employers exploit migrants, especially those without documentation (Bailey, 1987).

A related group of migrants come legally, with professional skills that are valuable to the United States. These migrants can enter with "skill visas" and often take up high-paying jobs. Most of these individuals come from a relatively few sending countries. In 1995 the United States admitted 110,000 persons classified as professionals and managers (Immigration and Naturalization Service, 1996); more than half were from China, the Philippines, and India. Most of the professional migrants do not come because they are unemployed or underemployed in their country of origin; rather, they migrate to increase their professional opportunities and are clearly intent on improving their human capital. These immigrants may enter at the lower rungs of the economic ladder, but their ladder extends to considerable heights, sometimes in their professional specialty. Initially their incomes may be modest, but they increase rapidly. Almost always their incomes are substantially greater than that of the foreign-born population as a whole.

Immigrant entrepreneurs are not a large group, but they play an important role in the new society because they often create jobs for their fellow migrants (Light and Bhachu, 1993). The concentration of entrepreneurs in Koreatown in Los Angeles and the wide variety of ethnic restaurants throughout the United States illustrate the role of this group of immigrants, who have found a way to penetrate the host economy. Again, they represent a combination of human capital and family network effects on migration. Entrepreneurship, which is usually measured as some form of self-employment, is the main way in which Asian Americans adapt and create their social mobility. Korean immigrants have been particularly adept at creating self-employment opportunities; in 1990 more than one-third of all Korean immigrants were self-employed (Cheng and Yang, 1996). Along with the migrant professionals, the entrepreneurs are viewed as the groups who are "making it" in America (Portes and Rumbaut, 1996).

In contrast to those who migrate to increase their economic opportunities, the political refugees and asylum seekers are often the least successful immigrants. Refugees are certainly not part of the flow generated by chances of improving human capital or (at least initially) by the outcome of networks. Rather, they are immigrants because of political decisions made by nation-states. The relationship between the United States and the country of origin determines whether these immigrants will be considered refugees. The origin of refugee flows to the United States has changed over time. In the 1970s they originated mainly in Cuba and Vietnam; in the 1990s the major flows have come from the former Soviet Union. In 1995 there were more than 190,000 refugees from the Ukraine, as well as continuing flows from Southeast Asia (Immigration and Naturalization Service, 1996). Still, refugees are only about 10% of all immigrants legally admitted in that year

The flows of immigrants have changed over time, reflecting economic conditions and changing political situations. This changing context is important for understanding contemporary migration.

MIGRATION IN CHANGING SOCIAL AND ECONOMIC CONTEXTS

Because migration occurs at specific times, it is related to the changing economic structures of nations, and to the demography of the sending and receiving regions. Migration is also affected by changing technology and the flows of information. The mix of migrants arriving in California and their future trajectory in the California and U.S. economies are not the same as the mix and trajectory of earlier waves of migrants from Europe, who entered a different labor market and a different social context.

Migrants are coming to changing labor markets.
Migrants arriving in the United States and California in the last two decades have encountered very different economic circumstances than those who came in earlier decades. They are arriving during a fundamental restructuring of the economies of the developed world. The postindustrial developed world is based on technological innovation and vast service industries; it is no longer a world of manufacturing offering unskilled factory jobs (James, 1995). The jobs that immigrant groups filled in the first decades of this century have vanished or are vanishing. Immigrants used these jobs as a way up and out; they offered both security and the chance of occupational advancement to formally unqualified individuals from the working class. As manufacturing jobs have declined, they have been replaced by less stable, and often low-paying, service jobs.

By creating a demand for low-skilled service workers, the restructured U.S. economy has generated a situation in which immigrants can compete with native-born low-skilled labor. Insofar as the new immigrants are willing to work for less-than-minimum wages, they supplement, or even supplant, native-born workers. Immigrants, especially undocumented migrants, can furnish an easily managed labor force that provides services at relatively low cost to an advanced technical society. Jobs do not appear to be lacking. Immigrants are finding work, but they are finding work at the very bottom of an economic ladder that no longer has the same rungs to success as earlier waves of migrants used to move up the socioeconomic scale. In California, for example, one-quarter of all new Guatemalan and Salvadoran immigrants work as janitors, maids, or cooks (U.S. Bureau of the Census, 1992; Waldinger and Bozorgmehr, 1996).

The growth at the top and the bottom of the economic scale has generated a "dual economy" in which two increasingly separate realms exist in an increasingly segregated urban space. A segregated social system contains more rich people and more poor, with a shrinking middle class.[2] The "new" immigrants are flowing into this world. Thus the decline in manufacturing jobs, and the increase in new jobs requiring high levels of education and service jobs requiring few or no skills, have left those at lower educational levels (generally native-born Hispanics and blacks) in a marginal position in the new labor market.

The change from the production of goods to the provision of services is particularly notable in America's metropolitan areas. While jobs in manufacturing and construction have declined, employment in communications, finance, trade, and the numerous service activities of large cities has increased significantly. In California, the job changes were not simply a restructuring: manufacturing continued to grow, but the growth in manufacturing jobs has been outstripped by the growth in service jobs. In California, 22% of all jobs were in manufacturing in 1970, in contrast to 17% in 1990; and this proportion has certainly declined futher in the 1990s. In the United States as a whole and in California in particular, the major function of the economy is assembling, processing, and trading information, not goods and raw materials. At the same time, the growth in the service economy is bifurcated: this economy creates a demand for both high- and low-skilled workers (Waldinger and Bozorgmehr, 1996). Some workers enjoy stable jobs and high wages, while others are faced with temporary, part-time, low-paying menial labor.

Finally, the migrants are arriving at a time when average wages have been in a long period of decline (Figure 1.1). Until about 1972 or 1973 the labor market continued to reward American workers with ever higher wages. Since that time the overall trend has been downward, to levels similar to those of the late 1950s. In addition, wage inequality has grown: the wages of skilled workers and those with a college education have continued to increase, while the wages of the unskilled have declined substantially.

Migrants are arriving in a different social context.
The social context has changed as much as the economic milieu. Although

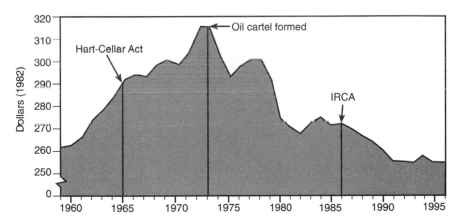

FIGURE 1.1. Changes in the weekly private wage in the United States. The Hart-Cellar Act was the change in immigration quotas and the IRCA was the Immigration Reform and Control Act. *Source:* Joint Economic Committe, 1997.

the "melting pot" description of the earlier migration flows was almost certainly too generalized to be an accurate term for the immigrant-absorption process, it did express a general long-run tendency for immigrant groups to become integrated into the wider society. Even so, immigration also brought groups who fit only awkwardly into American society initially and who found "melting" extremely difficult. Nonwhite immigrants and those from Asian and African countries did not blend easily into the so-called American culture, and many of them suffered discrimination. There was no discussion of immigrant or ethnic rights, nor was there much ambivalence about insisting on English as the only language.

When immigrants arrived in the first decades of this century, they came with little expectation concerning their rights, either as individuals or as groups, although the Irish, Poles, and Italians quickly formed local support and lobbying groups. Nor could the immigrant groups easily return to their native countries. Now the immigrants arrive in a society with a complex set of laws to protect their "new freedoms"; in addition, many new laws and policies, including civil rights and voting rights legislation, are specifically designed to protect the power and influences of legally defined minorities, including Hispanics and Asians. Although the laws are written to protect and empower citizens, they have spillover effects on the noncitizens and perhaps even on illegal populations as well. In the late 20th century the rights of discrete insular groups are emphasized in an atmosphere very different from that of the 1910s and 1920s.

The recent flows have shifted discussions from the "melting pot" analogy to the "salad bowl" analogy, from the expectation of "becoming American" to the celebration of pluralism and multiculturalism. Such terms emphasize separate identity rather than integration, at just the time when the flows are more diverse. Asians and Hispanics made up only small proportions of earlier flows, but these groups now dominate the new immigrant flows; Asians in particular are viewed as "culturally different" from earlier groups. Some descriptions emphasize separate identity rather than integration, and raise issues of local conflict in states and communities with recent large-scale immigration. The issue will be complicated further by the 2000 U.S. Census, which will provide for multiple racial identification in a "check all that applies" approach to self identification. Earlier immigrant groups were also spatially and socially separate from other groups but that separation gradually dissolved. Will the new immigrants also meld into the larger society?

A related issue for studies of immigration is the changed status of the minority African-American population. African Americans occupied a truly marginal position in American society at the turn of the century, when the last great waves of immigrants arrived; now they are simultaneously marginalized in the central city and making progress in the suburbs. For at least the isolated central-city black population, the changed composition of flows has created indirect (if not direct) conflicts between the new immigrants and native-born blacks because the new groups only rarely identify with minority African Americans (Rose, 1993, p. 191). Indeed, Rose (1993, p.202) goes so far as to argue that the future dividing line, created by the increasing diversity in the immigrant flows, may be between non-black and black.

Migrants are moving in a changing political world.
The worldwide focus on immigration stems from the fundamental political changes occurring in Europe, the changes in Asia following the end of the war in Vietnam, and the extensive, prolonged economic recessions of the 1980s. As a result of these events, immigration outcomes in the United States can be regarded as part of a series of redistributive effects on a global scale. The reunification of Germany, the breakup of the Soviet Union, the restructuring of Europe, and continuing instability in the Middle East and in southern Asia have changed both political and social expectations. Some groups seek to create or to reformulate nation-states; others have become political refugees or have taken the opportunity to improve their economic position. Some of the sociopolitical changes can be traced to our "shrinking world." Information technology and the ease of international travel have combined to diminish the effective distances between populations and to increase the potential for contact. The sight of TV antennas outside the traditional houses of the African plains dwellers reminds us of how integrated our world has become.

At the same time, immigrants bring their ethnicity and their culture to the new society they enter, forming enclaves that interact in turn with national identity and citizenship. The modern nation-state, and certainly the United States, is defined by a constitution and a set of laws, which in turn are created by the people. Membership in the state is defined by citizenship. Most large nation-states have attempted—through citizenship itself, the organization of political institutions, and the use of language—to achieve cultural and political integration. It is the individual states in the United States that must respond to the social and political problems arising from increased ethnic diversity. Also, how do nation-states in the late 20th century, with its possibilities of increased global travel and residency in multiple locations, foster a loyal, unified population? Multiple citizenship may create divided loyalties and undermine "the cultural homogeny" that is the nationalist ideal (Castles and Miller, 1993). The recent remarks of activist groups who speak of "reconquering" California dramatize the tension between ethnic groups and the nation-state.

Migrants originate largely in the developing world.
A significant proportion of the migrants arriving in the United States and California come from countries of high fertility and rapid population growth. Even when fertility rates have declined, the potential for continuing growth is extremely high. Several developing nations are expected to double their populations in the next two decades. Mexico, the country of greatest concern in a discussion of Southern California, had a population of only 26 million in 1950. By 1996 the country had close to 96 million people. Even more critical, nearly 40% of the Mexican population is under 15 years of age. Estimates of the effect of a youthful fertile population range from an expected population of 110 million by the end of the 1990s to 150 million by the first quarter of the next century.

Other developing nations with links to the United States and Europe are following similar trajectories. For example, the Middle Eastern countries with the greatest potential impact on Europe include Egypt, which may grow from 55 million to 103 million in the next 30 or 35 years, and Iran, which may grow from 54 to 130 million. Even if these projections are halved, they create the context in which immigration will be defined in the next two or three decades. Immigration is not independent of these fundamental demographic changes.

The new immigrants are admitted to the United States but the greatest proportion move to California. This geographical concentration is creating much of the debate about immigration and immigrants.

THE FUTURE OF THE CALIFORNIA CAULDRON: THE CONCEPTION OF AN OPEN SOCIETY AND THE ROLE OF THE MELTING POT

The debates about the future structure of American society are encapsulated in the struggle with the change from a white/Anglo-dominated society to a more inclusive mix of peoples and races. Those who worry about the disuniting of America and the peaceful "invasions" (Bouvier, 1992; Schlesinger, 1992) are at odds with those who celebrate the increasing diversity of the United States and California (Portes and Rumbaut, 1996). The latter praise an open society and an open-door policy (Simon, 1989); others point to potentially long-term negative effects of so many new immigrants in such a short period. They believe that the flow of immigrants indeed is creating problems in the labor market and in society as a whole, as we become increasingly concerned with group rights rather than with the society overall.

The underlying tension pertains to the role of assimilation and the role of the new immigrants in 21st-century America. Will the societal fabric stretch to accommodate the new arrivals, will the new arrivals help to reform the social fabric, or will the fabric tear under the impact? The United States has received concentrated bursts of immigration in the past, followed by periods of little or no immigration. Thus, the "assimilation" or melting occurred slowly as the pot bubbled. Will the pot boil with the tensions of accommodating many different groups with different agendas? Some observers worry that the emphasis on diversity and pluralism will lead to cultural separatism and divisiveness, and that the exaggeration of ethnic differences will drive "ever deeper the awful wedges between races" (Schlesinger, 1992, pp. 81–82).

It is not clear, however, that ethnic/racial pluralism implies division and decomposition. The real issue is how pluralism is accommodated. If ethnic groups are empowered under a dominance mode, in which one group is assured majority status in some political unit, division and divisiveness may well result. Under a different mode, however, one in which several groups join to influence outcomes, the separateness of ethnic groups will work to empower the totality. Thus, the way in which pluralism is accommodated is as important as the confrontation between assimilation and pluralism. The recent emphasis on ethnicity and diversity, then, may be viewed as a long-overdue recognition of the role of minorities in the forging of American culture.

Despite the fears of division and divisiveness, there are indications that many of the newcomers are following a path carved out decades ago by European immigrants. Like those groups, many of the new immigrants have asserted their historical identities as well as following the earlier migrant patterns from cities to suburbs, and from the periphery of political involvement to the center. Like their earlier counterparts from Europe, they move in, move up and move over (Rose, 1993). In the past, however, English-speaking Americans did not have to assimilate; assimilation was the task of the non-English-speaking Europeans.

There is a fairly intense debate about just how it will all work out. Clearly, non-European groups have assimilated slowly, if at all and this lack of assimilation has raised questions about the effectiveness of assimilation for integrating different races and ethnicities into the larger American society. Following this line of thinking, Glazer (1993) comes down on the side of multiculturalism, of a society to be defined by the choices of the varying Asian, Middle Eastern, and African groups. He suggests that their support for bilingual education and for foreign-language rights suggests a tendency to multiculturalism and may even portend a resistance to assimilation.

Some individuals however, have intermarried and to that extent have truly assimilated; it is just that the assimilation process was neither rapid nor smooth, nor (as I suggested above), a complete "melting." Vastly different ethnic groups, although mostly European, have been assimilated into American society. The process was difficult, however, and it occurred over a long period of very low levels of immigration. Between 1925 and 1965, immigration was at much lower levels, perhaps a few thousands a year. And most certainly conflicts existed between ethnic groups then, as now. Bayor (1978) provides a telling picture of conflict between the Jews, the Germans, the Irish, and the Italians in New York City. In some sense we have reentered that earlier cauldron with different actors. The competition and interracial/interethnic tension are the same but the issue now is whether we will become further separated or whether we will look back on this period as the beginning of another great time of renewal and assimilation. If California is the test case, how the process is worked out here will have important implications for the rest of the nation.

To reiterate the basic concerns of this book and the analysis on which it is based is to ask, To what extent will the new immigrant groups follow the old pattern of "integration"? Or will a new model of pluralism be constructed, in which groups maintain their racial and ethnic identities but are integrated into the larger social and political society? The evidence thus far does not convincingly support either case. As we will see later, however, at least some Hispanics who move to suburban locations in Los Angeles County and to Orange County are more likely to be longer term residents, to speak English well, to be citizens, and to earn higher incomes. This fact certainly supports the notion that Hispanics may follow the pattern of earlier European migrants. At the same time, the measures of concentration show that after an initial decline in the levels of separation, the very large numbers of Hispanics may be creating resegregation. Thus a variant of the melting pot emerges, in which ethnic separatism and social integration occur simultaneously.

PREVIEW

While the themes of declining human capital and growing inequality between recent immigrants and the native-born population are part of some recent studies of migration, this book emphasizes the geographic nature of these themes. I draw out the regional and community patterns of these changes and examine the implications for California as a whole. I argue here that the growing spatial inequality (increasing differences between poor inner cities and affluent suburbs) is closely related to the arrival of large numbers of immigrants, who are not well equipped to perform in a postindustrial society, and who thus may be relegated, perhaps forever, to the lower rungs of the economic ladder. Many of those who may be unable to ascend the socioeconomic ladder are the children of recent immigrants, and they are as much a matter of concern as the flow of new immigrants.

Other studies have debated whether immigration is "good" or "bad." Here I examine what is happening to places where there are significant concentrations of new immigrants. The large, sustained increase due to new citizens' high fertility will have long-term impacts well beyond the recent large-scale flows. Even if the birth rate declines rapidly, the data show that there will be a very large number of children born to recent immigrants. How will these new citizens be educated to provide the educated workforce of the 21st century? The current data on education in California do not always tell a cheerful story; the information suggests the emergence of a potentially low-income and low-achieving population, most visibly in very selected communities. Even so, local variations and different paths of achievement and assimilation for different groups of immigrants create a complex picture of immigrant California.

California is the immigrant entry port. More immigrants come to California than to any other state. Moreover, while the economic differences between Mexico, the Southeast Asian nations, Africa, and California continue to be so great, the flows will continue. Along with large numbers of new immigrants, there is an increasing number of children who are poorly prepared for the U.S. education system. The number of limited English proficiency (LEP) students is increasing and today already makes up nearly one-quarter of all students in California schools. Moreover, the continuing high fertility rate will increase the numbers of school-age children. The process is in place and will affect local communities for the next three decades. Without changes in national immigration policies or redistribution of resources to educate the influx of new students, the outcomes will affect California and the United States for the rest of the next century.

Some immigrants are doing well. Their economic trajectory may not be as steep as that of the native-born white population, but they are working and many have made significant economic gains in their new home. At the same time, however, the immigrants from Mexico and Central America are not only at the bottom of the social and economic ladder; they are farther from the top than were earlier arrivals from the same geographical areas. They will find it even more difficult to move up the ladder in the coming decades. Yet, although many of the new immigrants have limited schooling or even no schooling, they do get jobs, but their future path is related directly to their human capital: their education and training.

There are economic successes. Many immigrants are succeeding in the "New California"; they are achieving the "American Dream" of home-ownership and a good education for their children. From the perspective of recent immigrants, coming to California offers chances not available in their home countries; for them, settlement in California is clearly an improvement. Although the notion of assimilation has been challenged, substantial numbers of new immigrants are "making it," and the data show that this is especially true for immigrants from the Middle East and from Asia, but again not equally, and not for all.

California is the cauldron in which immigration has created a plural rather than a dual society. Many of the new immigrants do not share much in the way of a common language, culture, or religion. They often live in separate enclaves, the rich as well as the poor. The region's residential areas give evidence of increasing segregation, especially for Hispanics. Are some immigrants and immigrant groups falling behind and deviating from the upward trajectory enjoyed by earlier waves of migrants? Do present-day trends portend an increasingly fragmented society in which groups will live out separate lives, linked tenuously through skill and education? How will the patterns of ethnic separation evolve in a context of ethnic pluralism, perhaps accelerated by the continuing out-migration of non-Hispanic whites? These are difficult questions to answer, but they are central to the future of a changing society. Will our society accommodate the changes, or will it fracture with an ensuing balkanization of neighborhoods and communities? The tentative conclusion is that time is the critical factor; without time to absorb the new immigrants, to educate and train these new citizens, we risk increasing fragmentation and separation.

At the end of the book I revisit a critical issue: the future path of immigrants and the kind of society they will remake in California (and eventually the nation). Generalization and prediction are difficult, but the transformation of new immigrants takes time; the last chapter evaluates the pluses and minuses of such a dramatic change in California society. Because California is projected to grow to nearly 50 million people in the next decade, how those new Californians mesh is a matter of more than academic interest. From today's vantage point, one can only sketch out alternative scenarios of future possibilities. At the same time, these scenarios are guides to the problems that will undoubtedly emerge in the evolution of a pluralistic society.

NOTES

1. This is not the first time migration has appeared on the national agenda. The size and nature of migration into the United States were national concerns in earlier periods. For example, a 100 years ago explicit laws were passed to exclude additional Chinese immigrants; during the 1920s, laws were enacted that effectively halted immigration for 3 decades.

2. A substantial literature has examined the declining proportion of earners in the middle-income range, but a specific California study showed that the proportions of both Hispanic and non-Hispanic white households in the $25,000 to $50,000 categories decreased between 1970 and 1990. Hispanic households in that range declined from 46% to 31%; non-Hispanic white households declined from 42% to 30%. At the same time, the proportions of the lowest and highest income categories increased (Clark and McNicholas, 1995).

CHAPTER 2

CALIFORNIA:

THE NATION'S PORT OF ENTRY

The increase in the size and diversity of recent migration to the United States and California was unexpected. "Boat people" from Cuba and refugees from Rwanda are only the most visible manifestations of the new flows. Immigration flows have increased dramatically in the past two decades. Between 1980 and 1996 nearly 15 million new immigrants, almost 1 million migrants a year, came to the United States.[1] The immigration flows from Asia and from Central and South America are now comparable to the flows from Europe to the East Coast in the first two decades of this century. The largest proportion of the new immigrants settle in California.

Any historical reconstruction emphasizes the large-scale immigration of the period from 1900 to 1924, the year in which the United States ended its "open door" immigration policy. During that period, approximately 18.5 million immigrants were admitted to the United States. In a comparable 24-year span, 1971–1995, the country received about 16.5 million new legal arrivals. The fluctuating size of the flows is an important aspect of their impact, but their changing composition is equally important. Europe dominated the flows in the 1900–1924 period (90% of all immigrants), while Latin America and Asia dominated the flows in the 1970s and 1980s (over 80% of all immigrants) . If undocumented migrants are added, the flows in the 1970s and 1980s are certainly larger than those in the earlier decades of the century.

The most recent flow of immigrants, since the Hart Cellar Act of 1965, changed the preference system to emphasize migration based on skills and on family reunification rather than national quotas (Lapham, 1993), is sometimes described as a "fourth wave" of immigrants. The first and second waves were the flows of immigrants after the founding of the nation and during the middle of the 19th century. The third and previously largest wave was the flow between 1880 and the Immigration Act of 1924, which heavily restricted the number of visas and also included a provision to maintain the current ethnic structure of the United States. Until 1924 the flow was largely European, with some immigrants from China and Japan, and provided much of the workforce for the expansion of the mining, steel-production, meat-packing, and machinery-fabrication industries in the United States. The third wave was followed by a trough created by the restrictive immigration policies of the 1920s, by the Great Depression, and by World War II. From 1925 to 1950 only about 1.5 million immigrants came to the United States.

The fourth wave is characterized not only by large-scale flows and non-European immigrants but also by a steady increase in the number of illegal entrants. In previous waves, undocumented migrants were a much smaller proportion of the immigrant flows. It is estimated that about 5 million undocumented migrants now live in the United States (Warren, 1997) and that the net flow of undocumented migrants is about 200,000 annually (Bouvier, 1998). The most recent estimates have revised that number downward to slightly more than 100,000 net new illegal immigrants a year (Mexico–United States Binational Migration Study, 1998).

Until the fourth wave began, international migration did not have very direct effects on California. California grew rapidly in the 1920s and again after World War II, but the growth was due to internal migration from the Midwest and, to a lesser extent, from the East Coast. This picture changed in the late 1960s, when flows of immigrants increased dramatically (Figure 2.1). Now the links to California are global in scale. More than 30 countries each have sent 10,000 or more immigrants to California in the last decade, and 18 countries each contributed more than 2,000 migrants in the past year, 1994 (Figure 2.2).

FLOWS AND DESTINATIONS

The 1965 reunification act changed the flows to California from a steady stream to a torrent of strangers knocking on America's door (Cose, 1992, p. 219). Moreover, immigrants tend increasingly to concentrate in California. In 1970, about one-fifth of all immigrants went to California, and 58% of all immigrants went to five large immigrant states. By 1990 more than one-third of all migrants were going to California, and 70% to the large five immigrant states (Table 2.1). Immigrants are going to places that already have large immigrant populations. Between 1985 and 1990, California received about three times as many immigrants as New York, the next largest destination state.

California is part of a band of immigrant entry ports that stretches

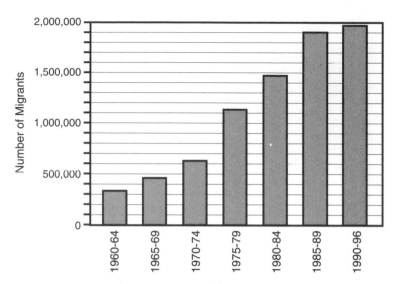

FIGURE 2.1. International migration to California by year of arrival. *Source:* U.S.
Bureau of the Census, 1992, and CPS, 1996.

from San Francisco to the Texas border at Brownsville (Figure 2.3). Five of
the 15 counties identified by the Center for Immigration Studies as immigration "hot spots" are in California, and 34 of California's counties were
identified as high- intensity-impact counties (Center for Immigration Studies, 1996). Although Los Angeles, San Francisco, and San Diego are obvious points of immigrant concentration, the effects of immigration extend
throughout Southern California and the agricultural communities of the
Central Valley.[2]

As a result of the recent flows, the foreign-born population in the
United States is now nearly 20 million people, and more than a fifth of
them live in California. California now has almost 8 million foreign-born

TABLE 2.1. Immigrants to the United States and 5 largest immigrant states (in 000's)

	1965–1970	1975–1980	1985–1990
California	356.8	1,111.5	1,906.6
New York	347.2	431.1	768.6
Florida	138.3	166.9	408.9
Illinois	104.1	196.3	232.4
Proportion in largest five	58.5	66.4	70.3
Total	1,721.1	3,295.6	5,275.2

Source: U.S. Bureau of the Census, 1972, 1983, 1992.

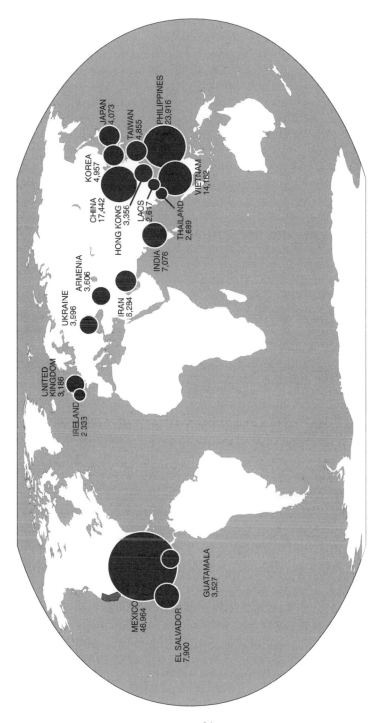

FIGURE 2.2. The source of legal immigration to California in 1994. *Source:* Immigration and Naturalization Service, 1995.

inhabitants, 27% of the state's total population (Figure 2.4), and almost as many as the other four largest destination states combined. The origins of the foreign-born in California are quite different from those for the United States as a whole, although California clearly exerts a massive effect on the U.S. statistics (Table 2.2). In 1990 the Latin American foreign-born population was a little less than half of all the foreign-born; by 1995 it was more than half. Although in 1990 the proportion of foreign-born of European origin in the state declined to about 10% of the total, that was still half of the European foreign-born population for the United States (Table 2.2).

The flow of immigrants to California has increased in every decade, and shows no sign of declining.[3] In fact, more migrants entered from 1991 to 1995 than from 1985 to1990. About 3.9 million new immigrants arrived between 1985 and 1995. These recent immigrants flows are dominated by moves from Mexico and Central America. At the same time, the total flows from the many Asian countries are rapidly approaching the size of those from Mexico (Figure 2.5). The flows from Japan have increased significantly in recent years, as have the flows from Southeast Asia (Vietnam, Laos, and Cambodia). These very large flows are defining the population composition in California in the middle of the last decade of the century.

The immigrants who entered California between 1985 and 1995 came overwhelmingly from Mexico. This finding is hardly new, but it deserves to

TABLE 2.2. Percentage distribution of foreign-born: California and United States, 1990 and 1995

| Place of birth | California | | U.S. |
	1990	1995	1990
Canada	2.5	1.6	4.0
Oceania	.7	.7	.5
Africa	1.1	1.7	1.7
Europe	11.1	9.7	23.6
Western Europe	8.4	5.9	17.8
Eastern Europe	2.7	3.8	5.8
Latin America	47.4	52.5	44.1
Mexico	36.7	41.1	23.3
Other Latin America	10.7	11.4	20.8
Asia	27.3	29.2	26.0
East Asia	10.3	11.1	9.1
South Asia	13.5	14.5	16.2
Middle East	3.5	3.6	.7

Source: U.S. Bureau of the Census, 1992 and Current Population Survey, 1996.

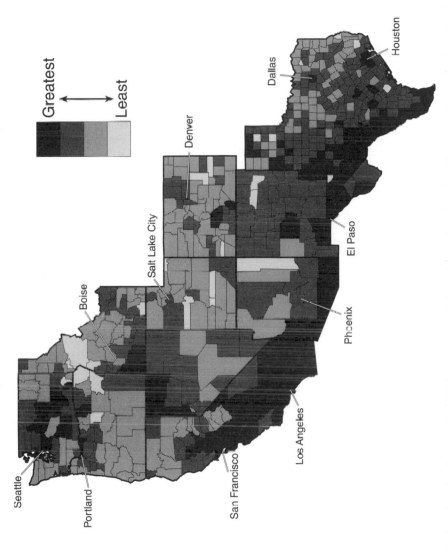

FIGURE 2.3. Immigration impacts in 10 Western states. Adapted from Immigration Hotspots, Center for Immigration Studies, 1996. reprinted by permission.

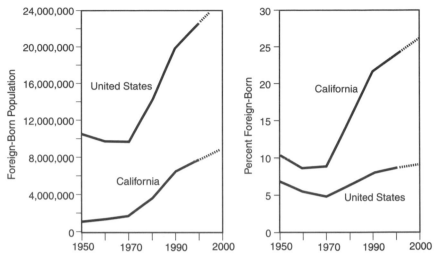

FIGURE 2.4. Total and percent of foreign-born population, 1950–1994. *Source:* U.S. Census of Population, 1950–1990, and CPS, 1994.

be seen in perspective. Although 40% of the migrants were from Mexico, an additional 32% came from Asian nations. The greatest number of the foreign-born are Mexican, and Latino generally, but recent arrivals from Asia account for flows that are almost as large. More than 1.5 million persons came from Mexico, almost 350,000 from Central America, and over 1 million from the East Asian nations. These numbers include legal arrivals and some undetermined number of illegal migrants. The composition of the migrant flows between the latter half of the 1980s and the first half of the 1990s did not change much.

The flows of 1970s and 1980s have changed the state's population composition dramatically. In 1995 California was about 54% non-Hispanic white (Figure 2.6). Whites are not yet a minority in the state, but the estimates from the U.S. Bureau of the Census suggest that the non-Hispanic white population will likely be a minority by the turn of the century. In contrast to today, in 1960 the population of California was 15.7 million, nearly 90% of whom were non-Hispanic white. In 1970 the state's population was just under 20 million; in the 2 decades between 1970 and 1990, it grew by 10 million—that is, by 50%. Between 1980 and 1995, the period of most recent intense immigration, the state's population increased from 23.7 to 31.2 million, including more than 1 million Hispanics and more than a half-million Asians (particularly Koreans, Filipinos, and Vietnamese). This change in composition has generated soul searching about the future of the United States and of California as an immigrant society.

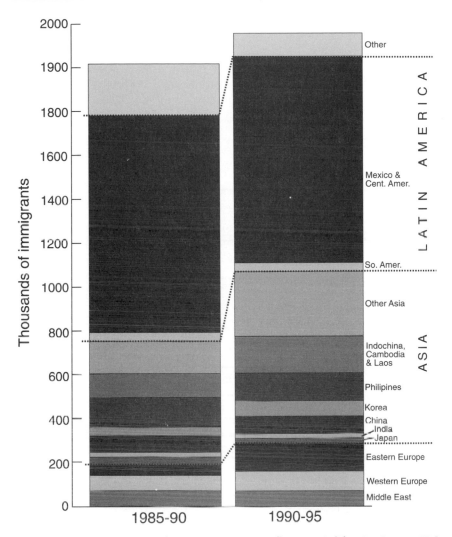

FIGURE 2.5. Composition of recent immigration flows to California. *Source:* U.S. Bureau of the Census, 1992, and CPS, 1996.

About 2.2 million legal immigrants were admitted to California in the period from 1984 to 1995 (Table 2.3). This number excludes the Immigration Reform and Control Act (IRCA) immigrants as reported by the Immigration and Naturalization Service and tabulated by the California State Department of Finance. The numbers have ranged from 150,000 to 200,000 each year, but have declined slightly in the last 2 years. The proportion of all migrants to the United States who were admitted to Califor-

26

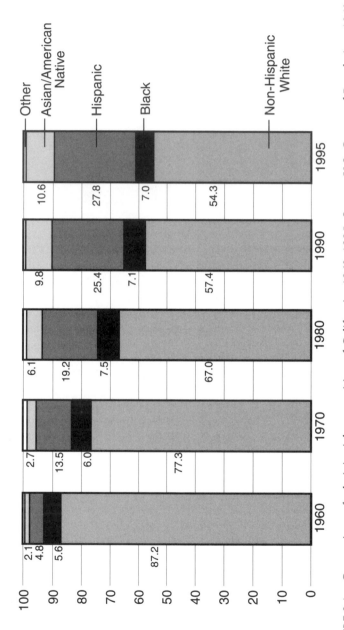

FIGURE 2.6. Comparison of ethnic/racial composition of California, 1960–1995. *Source:* U.S. Census of Population, 1960, 1970, 1980, and 1990, and U.S. Bureau of the Census Estimates, 1995.

TABLE 2.3. Legal immigrants (excluding IRCA legalizations) to the United States and California, 1984–1995

Year	U.S.	California	CA/U.S.(%)
1984	543,903	140,289	25.8
1985	570,009	155,403	27.3
1986	601,708	168,790	28.1
1987	601,516	161,164	26.8
1988	643,025	188,696	29.3
1989	612,110	180,930	29.6
1990	656,111	186,225	28.4
1991	704,005	194,317	27.6
1992	810,635	238,281	29.4
1993	880,014	247,253	28.1
1994	798,394	205,872	25.8
1995	720,261	166,482	23.1
Total	8,141,691	2,233,702	27.4

Source: Demographic Research Unit, State of California, 1997a,b.

nia ranged from 26 to 30% over this period (Demographic Research Unit, State of California, 1997a).

Many of new immigrants in the period 1984–1995 were processed under the 1986 ICRA, but 1.4 million of these immigrants were already present in California in 1982.[4] These individuals are not truly migrants in the sense of recent arrivals, though they are included in the 1990 U.S. Census and in the Current Population Reports. The effect of ICRA is quite dramatic and affected the immigrant totals through 1994 (Figure 2.7).

Of the 2.2 million non-IRCA immigrants, 60.3% were family sponsored. These "family reunifications" are a function of U.S. citizens and legal permanent residents, who sponsor family members; these family members are then admitted under country and numerical quotas. Family-sponsored migration, part of the process of chain migration, is self-perpetuating and can add very large numbers to the migrant population. An additional 24% of the immigrants were "noncapped" admissions, for whom no specific quotas are set. This category includes refugees, asylum seekers, children born after issuance of the entry visa to their parents, and other persons admitted for humanitarian reasons. Only 11.5% of all legal admissions were based on employment. Small groups were admitted for children born to alien residents and for legalization of dependents.

Legal immigrants came from 130 different countries, but predominantly from Asia (Table 2.4). Asians, mainly from the Philippines, Vietnam, Korea, and China, account for more than half of all legal immigrants. Mexicans are the single biggest group and together with Central

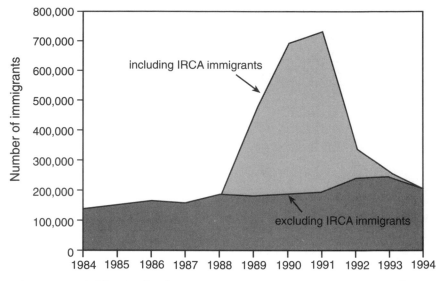

FIGURE 2.7.   California's legal immigrants, 1984–1994. *Source:* Demographic Research Unit, State of California, 1997a.

TABLE 2.4. Countries of birth for flow of more than 10,000 legal immigrants, 1984–1994

Latin America	Size	Asia	Size	Middle East	Size	Europe/ Canada	Size
Mexico	400,863	Phillippines	269,881	Iran	89,751	Former	44,964
El Salvador	79,005	Vietnam	192,280	Armenia	15,478	USSR	
Guatemala	30,811	China	124,866	Lebanon	13,238	UK	33,625
Nicaragua	19,641	Korea	89,612	Israel	10,383	Canada	23,499
Peru	14,237	Taiwan	63,594			Romania	13,409
		India	60,853			Germany	12,196
		Laos	40,723				
		Hong-Kong	35,517				
		Thailand	30,811				
		Cambodia	29,123				
		Japan	21,644				
		Pakistan	13,397				
		Indonesia	10,830				

Source: Demographic Research Unit, State of California, 1997a.

American arrivals account for about 27% of all legal migrants. The proportion of retirement-age migrants has increased (the effect of bringing in elderly family members) and the number of very young migrants has decreased slightly.

Forty-one percent of all immigrants to California settle in Los Angeles County, and 9% settle in Orange County. Another 20% report on their arrival forms that they intend to settle in the Bay Area. The concentration of new immigrants is even greater than suggested by the county-level figures. More than 16,000 immigrants settled in Zip Code 91205 in Glendale, another 15,000 in Zip Code 91754 in Monterey Park, and more than 25,000 in two Zip Codes in San Francisco, 94112 and 94110. Another five Zip Codes in Los Angeles contained more than 12,000 new immigrants in the period from 1984 to 1995. In contrast, 90210, perhaps the most famous Zip Code in California (it signifies the wealthy Beverly Hills community, and owes its fame to the TV program *Beverly Hills 90210*), had very few immigrants at all. These concentrations again emphasize the links to existing immigrant communities and the chain effects of earlier migration. Some Zip Codes are ethnically diverse (San Francisco's 94110 is home to Asian, Mexican, and Central American migrants), but the Glendale area consists almost totally of immigrants from Iran and Armenia (Allen and Turner, 1997).

Like immigrants in general, refugees and asylum seekers also settle disproportionately in California. About 30,000 arrive each year; over the period from 1990 to 1994, 65% of the refugees came from Asian countries. Vietnam sent almost 60,000 refugees between 1990 and 1994; Laos, Cambodia, and Thailand sent another 35,000. In total, 191,162 refugees arrived in California in the period 1990–1994. Seven counties in California (Los Angeles, Orange, Santa Clara, Sacramento, San Diego, San Francisco, and Fresno) received one-quarter of all refugees to the United States.

THE MEXICAN CONNECTION

Immigrants to the United States have diverse national origins but it is the link with Mexico that defines much of the immigration process in the late 20th century. The United States is the predominant destination of immigrants from Mexico and Central America; as we have seen, most of those immigrants settle in California. Between 1980 and 1995 more than 3 million people migrated from Mexico to the United States. Two million of those, nearly 20% of Mexico's net population growth, came to California. In 1995 about 2 to 4 million of Mexico's 30 million workers relied on the U.S. labor market for most of their annual earnings (Martin, 1995).

Mexico, now a country of 96 million people (Population Reference Bureau, 1997), has grown rapidly in the past 30 years (Table 2.5). As recently as 35 years ago the Mexican population was only 34 million. From 1970 to 1996 the population almost doubled, from 48 million to the current 95 million. Although population projections are susceptible to changes in fertility, to reiterate an earlier statement, the very large number of Mexican women aged 15 to 45, the critical childbearing years, suggests that the Mexican population could reach 141 million in the next 25 years (Population Reference Bureau, 1997). The age/sex pyramid for Mexico in 1990 shows many women of childbearing age, but even more important, the "bulge" of 5- to 15-year-old females in 1990 has reached, or soon will reach, childbearing age (Figure 2.8). The total fertility rate (essentially the average expected number of children per woman) for Mexico is 3.1. Even more telling, the number of births per 1,000 population is 27 for Mexico and 15 for the United States as a whole. The natural population increase of the Mexican population is 2.2% a year but only 0.6% a year for the United States. By any measures, the Mexican population has high fertility, is growing rapidly, and is adding very large increments to the working-age population. Even though the percentage under age 15 is declining, the number of children in 1997 was approximately 34 million.

Fertility rates in other Latin American countries are even higher. For example, Guatemala is expected to double in population from 11.7 million to 21 million by 2025, and El Salvador will increase from 7.3 to 9.1 million (Population Reference Bureau, 1997).

The growth in jobs in Mexico is not keeping pace with the increase in the population. Approximately 3 million new jobs were created in the past decade, while the working-age population grew by five times that number. The difference between population growth and job growth is a critical element of the migration dynamic. Thus, macroeconomic conditions play an important role in the migration process. Because the Mexican economy did not perform well in the 1980s, and because the population grew 2.4 to 2.7% per year during that decade, there is an underlying migration push,

TABLE 2.5. Mexican population by decade and percentage of the population under 15

Year	Total	Under 15	% under 15
1950	25,791,071	10,754,468	41.70
1960	34,923,129	15,452,107	44.24
1970	48,225,000	22,287,000	46.21
1980	67,382,581	28,856,895	42.82
1990	81,249,645	31,146,504	38.33
1996	94,800,000	34,128,000	36.00

Source: Census of Mexico, 1996.

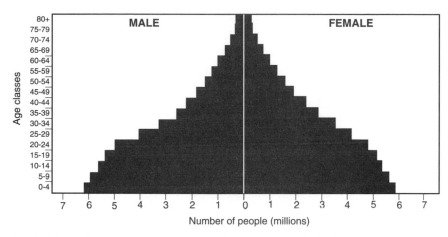

FIGURE 2.8.   Age and sex structure of Mexico. *Source:* U.S. Census Bureau, *International Data Base,* 1996.

especially from the poorer agricultural regions. In addition, growth in the formal sector fluctuated greatly. In 1984, the best year, the growth was 2.3 %. In 1986, however, growth was negative (1.7%), and in 1988 employment grew only about 1% (Navarro, 1994). Labor's share of national income declined from 43% in 1979 to about 30% in 1988.

The fall in real wages (Figure 2.9) and the formal unemployment rate of about 11% emphasize the economic motivation to migrate. In a 1989 survey of migrants from Jalisco to the United States, 58% reported that they had made their last trip to the United States because of unemployment or to earn higher wages (Arroyo, De Leon, and Valenzuela, 1990). If we include the 13% who reported that they migrated because they could not support their families in Mexico, almost 70% of the migrants reported economic "pushes" for their migration.

The devaluation of the peso, in 1982, in 1986–1987, and again in 1994–1995, may have influenced immigration, though not in any simple cause and effect relationship (Martin, 1995). Rather, the peso devaluation, along with increasing layoffs of urban workers, have slowly eroded the living standards of Mexican workers and have created additional pressure to migrate north. In addition, surveys emphasize the close and increasing links between Mexico and the United States; some surveys report that as many as half of all Mexicans have relatives in the United States and one-third have spent time in this country (Martin, 1995).

The potential for emigration from Mexico is substantial. The outcome of employment conditions in Mexico has created, and will continue to create, a stimulus to find work in the United States. The impact on California

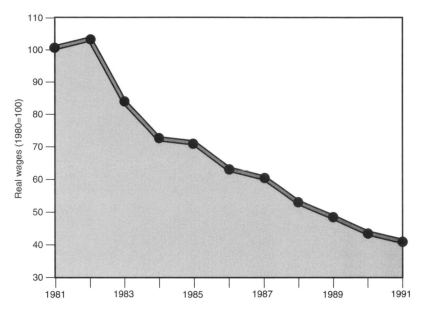

FIGURE 2.9.   Change in minimum real wage in Mexico, 1981–1991. *Source:* Economic Commission for Latin America and the Caribbean, United Nations, on the basis of figures provided by the Bank of Mexico and the Institute of Statistics, Geography, and Information Sciences.

will continue because this is the closest source of employment opportunities for the Mexican regions with high unemployment and growing populations: Jalisco, Michocacan, and Zacatecas. Because the border has been relatively porous, much of the job-seeking movement is undocumented.

Undocumented Migrants

The job and wage imbalance between Mexico and California, combined with the ease of movement across the border, has always encouraged undocumented workers to come to California (Jones, 1982). For the 1970s, the evidence indicates a pattern of circular migration in which Mexican seasonal workers journeyed north to provide supplementary incomes for rural households. In the 1980s, many of these workers began to stay, and their numbers increased. In the period before 1970, about one-third of the migrants had some form of legal document. By the early 1990s, only one-fifth possessed legal documents (Donato, 1994); the proportion with no documents jumped from 52% to 73%. In addition, the proportion of female migrants increased and the proportion working in agriculture declined.

Calculating the number of illegal immigrants in California is not a straightforward process. Early attempts to "count the uncountable" (Warren and Passel, 1987) focused on the differences between the reported census results in 1980 and alien registration data, the I-53 cards that all "aliens" were required to fill out during the 1960s and 1970s.[5] Those estimates have been reviewed and discussed in an extensive debate (Warren and Pasell, 1982; Woodrow and Passel, 1994; Warren, 1997). The panel on immigration statistics of the National Academy of Sciences concluded that between 2 and 4 million illegal immigrants lived in the United States around 1980 (Warren, 1997), and that approximately one-half of these were resident in California. The latest figures suggest that an illegal population of 5 million lived in the United States in October 1996 (Warren, 1997), including 2 million living in California.[6]

The Immigration and Naturalization Service (INS) estimates that the annual increase in the long-term illegal population (those who have been resident for more than a year) is about 400,000. When this figure is offset by deaths, changes in legal status, and out-migration, however, the permanent annual addition is nearer 275,000. The increase in the illegal population is due to both entry without inspection (EWI) and overstaying by legal entrants who have a tourist or other visitor visa. The illegal population is concentrated in the five high-immigrant states; 40% are in California. Fifty-four percent come from Mexico, again emphasizing the effects of proximity on California. The numbers represent a count of the long-term undocumented population and do not count short-term illegal entrants who stay less than a year. Short-term temporary illegal immigration is discussed by Van Hook and Bean (1998), but it is difficult to gauge this number with any certainty.

The latest numbers suggest that the net permanent additions to the Mexican illegal population are about 105,000 per year (Mexico-United States Bi-National Migration Study, 1998).[7] The Bi-National Study recognizes that illegal migrants do not travel only one way; many migrants return home, although the study also found that even those who return home once a year are tending to stay longer in the United States. The numbers do not conflict with the Warren (1997) estimates, which are of total undocumented migrants in the United States. These calculations do not take into consideration that many undocumented migrants marry and have children in the United States, and thereby become "anchored" in this country. The exact number cannot be known precisely; the numbers will continue to fluctuate with changes in policy and the two countries' economies.

Some sense of the incentive for illegal entry is indicated by the pent-up demand for visas. We might even view the waiting list for immigrant visas as a rough barometer of the possibility for further undocumented migration. In January 1997, the U.S. State Department reported that 3.6 million persons were "wait-listed" for visas. The waiting list consists of individuals who have filed visa petitions but who cannot be accommodated because of the limits on applications from particular countries. The waiting list is topped by Mexico and the Philippines (Table 2.6) and consists primarily of family-related applicants—that is, relatives of people already legally in the United States. Only 2.3% of all wait list applications were related to employment (Vaughan, 1997).

The demand for visas and the illegal entry from Mexico persist despite the attempt to control illegal immigration by enacting a significant change to immigration policy. That change, the Immigration Reform and Control Act (IRCA), was intended to change fundamentally the flood of undocumented immigrants who were being apprehended at the border in the early 1980s (Baker, 1990). The evidence, which is reviewed in the following section, suggests that illegal migration was not slowed; that it is concentrated heavily in California, particularly in Southern California; that it is having negative effects on the wages of earlier groups of Hispanic migrants; and that it will continue to enlarge the immigrant population of the border states, particularly California.

THE IMMIGRATION REFORM AND CONTROL ACT AND ITS AFTERMATH

The Immigration Reform and Control Act (IRCA) was an attempt to take charge of the immigration process. The IRCA provided amnesty for all immigrants who were in the United States before 1982, and at the same time

TABLE 2.6. Visa waiting lists for the United States, 1997

Nationality	Number	Category	Number
Mexico	1,020,823	Family	
Philippines	573,414	First	93,376
India	343,159	Second	1,630,621
China	235,175	Third	312,200
Dominican Republic	150,596	Fourth	1,502,233
Taiwan	108,625		
South Korea	77,203	Employment	84,467
Vietnam	75,568		
El Salvador	69,809		
Haiti	69,221		
Other	899,304		
Total	3,622,897		3,622,897

Source: Vaughan, 1997.

it created sanctions for employers who knowingly hired undocumented workers. It also allowed for a special legalization of agricultural workers, who had only to show that they had been employed for more than 90 days in the United States during 1985–1986.

As part of the IRCA, more than 3 million applications were filed under the IRCA program, including 1.3 million seasonal agricultural workers (Hoeffer, 1989, 1991). It has been estimated that as many as half of the persons legalized under the farm worker legalization programs were not truly eligible, according to the amnesty program (Martin, 1995). There were 1.63 million applications in California; 673,948 of these were filed by special agricultural workers. About 97% of the applicants were from Mexico and Central America, and were concentrated in Southern California. Seventy-five percent of all the legalization applications were filed in the five-county region around Los Angles and San Diego (Figure 2.10). The applicants were largely young (almost 40% were under 25 years of age), more than half were unmarried, and most were unskilled; only 43,000 of the applicants who had been here before 1982 worked in professional or executive occupations.

Initial examinations of the effect of the IRCA suggested that it had reduced the numbers of illegal border crossings. More recent studies, however, have concluded that the IRCA had little or no effect on illegal migration, which has continued to rise in spite of the IRCA (Bean, Edmonston, and Passel, 1990, p. 208). Moreover, others have noted that "the few small effects [of the IRCA] we have uncovered are little to show for the millions of dollars and the thousands of hours that IRCA has invested in an effort to stem the tide of Mexican immigrants to the United States" (Donato, Durand, and Massey, 1992, p. 56). That immigration would have been even higher without the IRCA is suggested by Espenshade (1995), who evaluated both the overall upward trend and the nature of the flow of undocumented aliens. These factors support the earlier argument that the potential for large-scale flows across the border continues.

As I noted above, the legalization applicants are concentrated in Southern California. Within that region, they are concentrated near existing stocks of immigrants. Strong concentrations of Mexican legalization applications are found in East Los Angeles in Los Angeles County and in Santa Ana in Orange County (Figure 2.11). Even though Zip Codes are not the ideal way to measure the locations of the "undocumented" population, some central Zip Codes contained very large numbers of applicants. Given the continuing surge of migrants, it is almost certain that these same areas will continue to be the ports of entry for new waves of undocumented migrants.

Evidence also shows that the legalization program depressed the

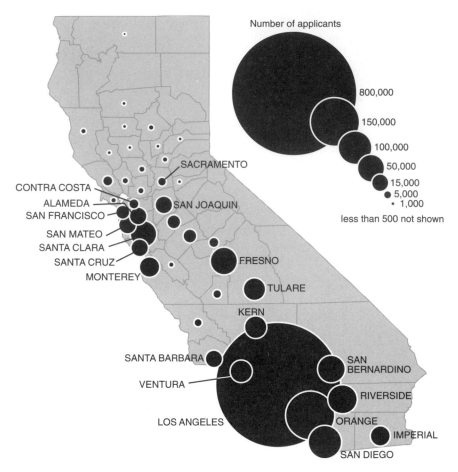

FIGURE 2.10. Total applicants for legalization in 1991 with residence before 1982. *Source:* Hoefer, 1991.

wages of other immigrant groups who already lived in California. The real hourly pay of Mexican immigrants with more than 10 years' residence in the United States declined by 13% from 1983 to 1989 (Sorenson and Bean, 1994). It is reasonable to expect that a sudden increase in the supply of new "legal" residents would depress wages in low-skill jobs. With greater competition for jobs, employers can pay lower wages. That the IRCA did not affect the wages of the most recent immigrants is also understandable because a pool of low-wage immigrants already existed. The critical effect was to slow the progress of longer term low-wage immigrants. The effects on other groups, including native-born Mexican Americans, appear to be

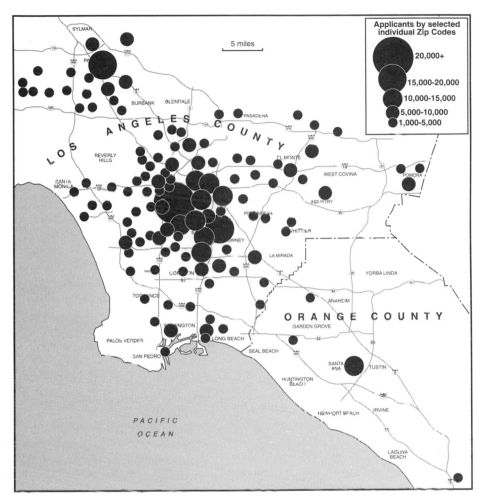

FIGURE 2.11. The concentration of Mexican-origin applicants for legalization under IRCA. *Source:* Hoefer, 1991.

negligible, although opinions vary on this issue and on whether the wages of native-born groups in low-skill occupations have beene affected (Smith and Edmonston, 1997).

The greatest long-term impact of the IRCA may not be on the immigration process itself, but on the increase in the number of children who become legal residents of California. The state has 766,203 citizen children who have foreign-born parents, and another 297,553 children who are not citizens and who have foreign-born parents (PUMS, 1990). The status of many of the citizen children, as well as the children who are not yet citizens, is related to the extension of amnesty as part of the IRCA program. Initially, they were children of illegal residents. A recent small survey estimates that one-quarter of the children born in Los Angeles County were born to undocumented residents (Heer, 1998).

OBSERVATIONS

As long as strong differences exist between the economic opportunities in Mexico, the Philippines, North Korea, and China, on the one hand, and those in California, on the other, we can expect a continued flow of immigrants into the state. The research shows that the flow of migrants is particularly susceptible to changes in the Mexican economy. When the peso is devalued or when Mexican wages decline, the result is an appreciable increase in the number of border arrests for illegal crossings (Martin, 1995). It is not only economic differences, however, that are fueling the increase in migration. The population of Mexico, especially the youthful population, is growing. The increase in the working-age population far outstrips the ability of the Mexican economy to provide jobs for these new entrants to the labor market. But across the border in California there are plenty of jobs, even though they pay poorly.

We might argue that immigration policy is a central issue for global social well-being, just as environmental policy is central to global ecological well-being. In this connection Layard, Blanchard, Dornbush and Krugman (1992) from the perspective of Eastern and Western Europe, argue that "rich" countries must provide at least conditional aid within their regions. From their economic perspective, trade, as much as flows of capital and people, will help living standards in developing and developed countries to converge. Absent the recognition of the disparities and without a regional perspective, the immigration pressures will rise and the flows will reach unsustainable levels.

History tells us that walls designed to keep people out have not been successful. We see no indication that barriers will be any more successful in the late 20th century than they were in the past. A comparison of migration pressures in Mexico, Morocco, and Turkey suggests that it will be much more difficult for California to manage migration from Mexico than for the European Union to manage migration from Morocco and Turkey. The demand for cheap labor in a largely uncontrolled labor market will continue to attract new migrants, and clandestine migration will continue as older clandestine workers receive amnesty (Bustamante, Reynolds, and Hinojosa-Ojeda, 1992; Muus, 1996). Thus far, Southern California has provided evidence that neither barriers nor changes in the immigration program succeed in preventing a continuing stream of undocumented migrants to the metropolitan areas of the United States.

An ancillary issue is raised by large-scale and concentrated undocumented migration. About one-third of all immigrants settle in California; about three-fifths of these new immigrants come to the Los Angeles metropolitan area. This concentration, more than any other issue, raises the question of the nature of a nation-state and the "rule of law." If the patterns of admission are different for some groups than for others, the nations' view of itself may be undermined. This is both the most amorphous and the most critical issue of the debate about migration; it raises questions of how entry will be decided and how assimilation will proceed.

NOTES

1. There is a question of exactly how a migrant is identified. Two different questions are asked, "When did you come to stay" and "Where did you live 5 years previously?" Ellis and Wright (1998) have suggested that the actual number of recent immigrants may be overstated by asking the question "When did you come to stay"?
2. The high-intensity immigration "hot spots" are essentially those communities where the number of immigrants is large and where they make up a very large proportion of the total population. The index is composed of eight elements that measure the percentage of the under-18 foreign-born population as well as other variables such as the ability to speak English, those who speak another language at home, and the percentage who are recent immigrants (Center for Immigration Studies, 1996).
3. The numbers calculated from the Public Use Microdata Samples (PUMS) and from the Current Population Survey (CPS), 1996, overlap slightly. The data from PUMS are drawn from the census that was taken on 1 April 1990, and can include a small number of migrants from 1990 (those who entered in the first 3 months). The CPS data can include the same migrants because it asks for years of entry beginning in 1990.
4. The inclusion of the formerly illegal migrants produces a bulge in immigration. This bulge gives a misleading impression of recent immigration and raises California's share of all migrants to almost 30% of the national total because a very large proportion of the ICRA immigrants were California residents.
5. The procedure compared the legally resident alien population with the 1980s census counts of foreign-born persons, corrected for misreporting of citizenship and country of birth. The estimate of the undocumented is the difference between the count from the Immigration and Naturalization Service reporting and from the Census Bureau reporting, plus an estimate of the illegal immigrants missed in the 1980 Census. The method is discussed in detail in a series of papers by Warren and Passell (1987).

6. Recent estimates of the undocumented population are probably about as reliable as can be achieved given the population's fluid nature, its aversion to contact with government agencies, and its fears of repatriation to the county of origin. The 1985 estimates have been updated, but one must recognize that the illegal population is not constant, either in size or membership; rather, it has tended to be a procession of people who come and go. The estimates suggest that the illegal population peaked at 5 million in 1986 and then declined to approximately 2.8 million at the end of 1988, after 3 million applied for legalization in 1987–1988 (Warren, 1997).

7. There are almost certainly issues of accuracy in the estimates. There is the possibility of both underestimates for some groups and overestimates for others (Camareta, 1997). Even so, the numbers have been reviewed carefully and provide the best current estimate of the size of the illegal population in California: about 6.5% of California's total population. At the same time, there have been lively debates about the state level accuracy of the estimates of the undocumented (Woodrow-Lafield, 1994). This problem may be smaller in California, where the numbers of immigrants are very large, than in states with smaller numbers of immigrants. The most recent analyses have reduced even further the size of the illegal flows.

CHAPTER 3

THE STATE IN FLUX:

COUNTY AND LOCALITY EXPERIENCES

I n the previous chapter I examined the flows and the changing intensity of immigration to California. Here I discuss the aggregate consequences of immigration, both direct and indirect. I consider how the state as a whole and particular localities are faring under the changes due to immigration and why the local fortunes of California's communities[1] vary.

MIGRATION PATTERNS OF THE FOREIGN-BORN

The foreign-born population is not distributed evenly across California's counties and communities; it is concentrated in the Bay Area of Northern California and in Los Angeles and Orange Counties in the southern part of the state (Figure 3.1). Los Angeles and San Francisco Counties are the only counties where more than one-third of the population is foreign-born. Other coastal and agricultural counties have foreign-born proportions of about 20%. Given the size of recent flows, it is not surprising to find counties where the recently arrived foreign-born make up over 50% of the foreign born population, especially in the agricultural counties of the Central Valley of California (Figure 3.1). Thus many of the foreign-born are new arrivals and are not yet citizens.

Los Angeles County is the major destination of recent immigrants. A map of immigrants who arrived between 1985 and 1990 shows both the dominance of Los Angeles County and the split between Mexican and other migrants (Figure 3.2). Southern California, however, is not the only destination for large numbers of Mexican migrants; San Joaquin and Fresno Counties, with their large agricultural bases, also attract low-wage Mexican labor to pick and harvest the local field crops. The north–south split in the size and the composition of the immigrant flows has implications for

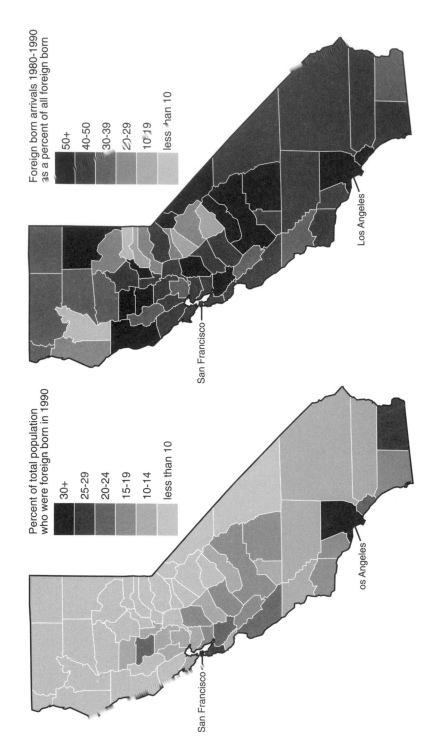

Percent of total population
who were foreign born in 1990

30+
25-29
20-24
15-19
10-14
less than 10

San Francisco

os Angeles

Foreign born arrivals 1980-1990
as a percent of all foreign born

50+
40-50
30-39
20-29
10-19
less than 10

San Francisco

Los Angeles

FIGURE 3.I.   Foreign-born and recent foreign-born by county in 1990. *Source*: U.S. Bureau of the Census, 1993.

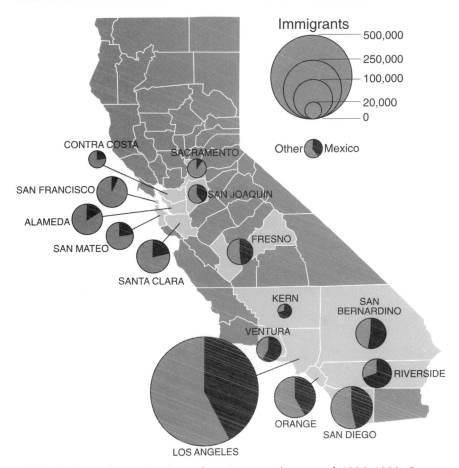

FIGURE 3.2. Major destinations of immigrants who entered 1985–1990. *Source:* Clark, 1996. reprinted by permission of Carfax Publishing.

education, poverty, and dependency, and is an important distinguishing characteristic of the patterns of migration in California.

STATEWIDE IMPACTS

Fertility

The new migrants are young, like other waves of migrants before them, though the immigrants from Mexico and Central America are much younger than those from other regions. All the immigrant age/sex pyramids reveal a bulge in the younger age groups, but this bulge is most pronounced in the age pyramids for the 1985–1990 and 1990–1995 Mexican/Central American migrants (Figure 3.3). The bulge in the 20- to 34-year-old age group is dramatic for the Latino immigrants. We also see an interesting reversal between 1985–1990 and 1990–1995: male migrants dominated the first interval, but female immigrants dominated the second one. This is probably an outcome of legalizing many young men during the IRCA, who then were able to marry and bring in their new spouses. Asian

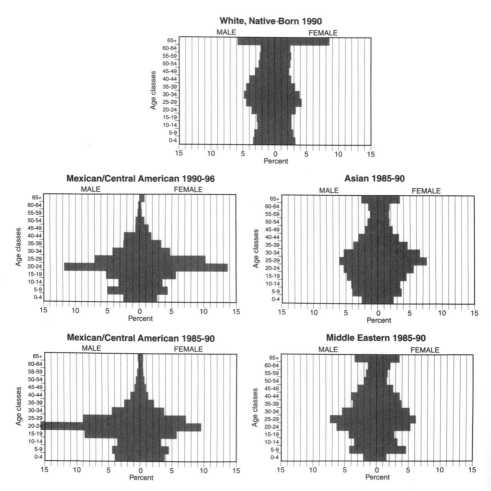

FIGURE 3.3.   Age/sex pyramid for selected immigrant populations who migrated to California 1985–1990 and for California non-Hispanic white population. *Source:* U.S. Bureau of the Census, 1992.

and Middle Eastern immigrants included a significant proportion of elderly persons; the spread at the top of the age pyramid is at least partially due to the family reunification policy, but in these cases for a different age group. In this instance, younger Asian and Middle Eastern immigrants brought in their elderly parents.

In contrast, the native-born non-Hispanic white population in 1990 resembles a rectangle more than a pyramid, and is dominated by the very large proportion of the white population that is now over age 65. The most vivid way to show the youthfulness of immigrant populations is by comparing the aging white population with the extremely youthful Hmong population (Figure 3.4). The potential mothers in the "bulge" were children under age 10 in 1990, but we know that they are likely to have large families and to add significantly to the total California population. They will add fewer children than the Mexican population, however, even if their birth rate is higher, because there are fewer Hmong than Mexicans living in California. Therefore two factors contribute to the increase in children, and thus to the overall increase in the population: the number of women of childbearing age and their fertility rate. The age/sex pyramids are a picture of future fertility because they depict the size of the population that will produce the future citizen children.

We know that fertility is significantly greater for foreign-born women in California than for native-born non-Hispanic white women. The latter have between 2 and 2.5 children, depending on whether they are in the labor force or not. In contrast, foreign-born Asian women and Hispanic women have between 3 and 4.7 children (Figure 3.5). The number of children born is also related to educational attainment: foreign-born Hispanic women with less than a 9th-grade education average more than 4.7 chil-

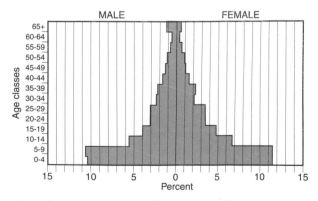

FIGURE 3.4. Population pyramid for Hmong in California, 1990. *Source:* Pacific Rim Research Program, University of California.

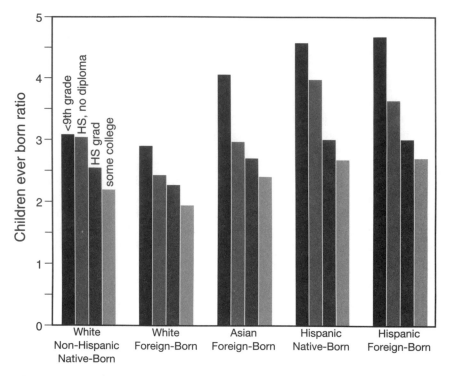

FIGURE 3.5.   Birth rates by native-born/foreign-born and educational attainment. *Source:* Heim and Austin, 1995.

dren during their childbearing years. In contrast, the fertility rate of college-educated foreign-born women is much closer to that of native-born white women: about 2.5 children if they are Hispanic, but fewer than 2.0 if they are non-Hispanic white. Hispanic native-born women with low educational levels do not differ very much from Hispanic foreign-born women at the same levels.

The outcome of differential fertility has already increased the number of children born to foreign-born mothers. In the last 2 decades, a period of accelerating immigration, births to foreign-born mothers also accelerated. In 1993 almost 45% of all births in California were to mothers born outside the United States (Table 3.1), and almost 75% of Hispanic births were to immigrant women. To reiterate, this pattern of increased childbearing by women of Mexican ancestry has important consequences for the continued growth of California's population.[2]

Even if immigrants follow the usual path of having families that are about midway in size between those of the original country and those of

TABLE 3.1. Annual births in California by birthplace of mother, 1970–1996

Year	Total	U.S.-born	%	Foreign-born	%
1970	362,652	324,375	89.4	38,277	10.6
1975	324,949	242,460	74.6	82,489	25.4
1980	402,720	286,873	71.2	115,847	28.8
1985	470,816	319,204	67.8	115,612	32.2
1990	611,666	361,688	59.1	250,278	40.9
1993	584,483	322,810	55.2	261,673	44.8
1994	567,034				
1995	551,226	306,364	55.6	244,862	44.4
1996	538,628				

Source: Demographic Research Unit, State of California, 1997.

the country of destination (Burke, 1995), the California Hispanic ethnic population will increase significantly. Because many of the immigrant women come from rural areas, where the completed family contains 5 or 6 children, the size of the families may not decrease as much. The data in Figure 3.3 show quite clearly that the number of Hispanic women of childbearing age will continue to grow for at least the next 2 decades. The very large increase in Hispanic immigrant women with relatively low levels of education and high fertility will have major effects on the size of the youthful California population. Those young people are the future of California, and how we educate and nurture them is a critical dimension of the state's future.

The other part of the equation is the increase in the Mexican population. As Mexico's fertility rate slows, the pressure for migration may decrease. Certainly, the difference in Mexican fertility between 1970 and 1990 is remarkable. In 1970 the average Mexican woman had approximately 7 children during her childbearing years; in 1990 the lifetime birth expectancy was closer to 3 children. Yet although the annual decline in fertility has slowed considerably, and may hover around 3 children per woman, it is considerably higher than the overall fertility in the United States, and will have the effect of continuing Mexico's population growth. Some observers expect the fertility rate to level off at about 2.5 children per woman (Haub, 1997), but the timing of the decline is not yet clear.

As I noted earlier, Central American countries have not experienced a similar drop in fertility. This may be part of the explanation for continuing high fertility rates among Hispanic foreign-born women in California. The average woman in Honduras and Guatemala still has 5 children, and a typical Nicaraguan woman can expect to have 4 children. Moreover, family planning has been much less successful in these overwhelmingly Catholic countries, where local Roman Catholic leaders

strongly oppose family planning programs, and where governments do
not have the money to invest in such programs. Overall, the trajectory of
fertility is critical for the future growth of the nearby immigrant origins,
and it will affect the future growth of population in California. At the
same time, the very large numbers of foreign-born women will change
their fertility behavior only slowly, and California's population will continue to grow because of internal fertility. Between 1975 and 1995 the
Hispanic share of the births in California, more than doubled from 20%
to 46% (Pinal and Singer, 1997).

An additional dimension of the high fertility is the large number of
teenage births. Although this number has declined in the very recent past
(Demographic Research Unit, State of California, 1997a), teenage births
have surged in the last decade and are much higher than those for earlier
waves of migration. Predictions suggest a continuation in substantial numbers of these births. By the year 2006 they are projected to rise to 90,000
annually (Figure 3.6), a projection that if accurate would account for one-
quarter of all births in California. As a result, the California teenage birth
rate is still one of the highest in the nation: for every 1,000 California teens
it is estimated that 154 become pregnant, well above the national rate of
111. As most of the Hispanic teen pregnancies come to term, it boosts the
proportion of Hispanic births. Almost two-thirds of the births are to His-

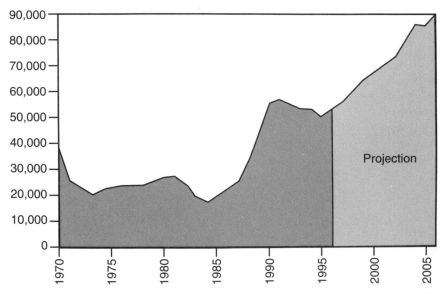

FIGURE 3.6. California teen births, ages 15–19. *Source:* Department of Health
Services, State of California, 1995.

panic mothers (Department of Health Services, state of California, 1995), and either to immigrants or to the children of immigrants.

Early childbirth interferes with schooling and preparations to enter the labor market, and thereby curtails the formation of human capital for the state as a whole. Teenagers who give birth are less likely than their non-parent peers to complete high school, and males who father children as teenagers are only half as likely to graduate from college as their peers who delay fatherhood (California Department of Health Services, 1997). Overall, teenage mothers earn about half the lifetime income of women who first give birth in their 20s.

Education and Skills

The new immigrants are young, and often unskilled when they arrive (Figure 3.7), but we cannot simply place all immigrants into the same category. Whereas 76% of the Mexican immigrants have less than a high school education, and almost none have a college-level education, the Asian (other than Southeast Asian) and Middle Eastern immigrants are well educated

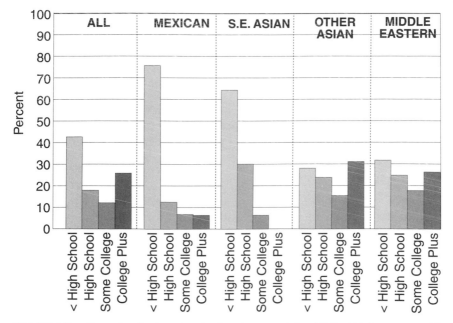

FIGURE 3.7. Completed education by immigrant origin, 1990–1996, except Middle Eastern, 1985–1990. *Source:* U.S. Bureau of the Census, 1992, and CPS, 1996.

and many have college training. The Southeast Asian migrants, often refugees, are more like the Mexicans than like the other Asian migrants. Because of their relatively low levels of training, it is not surprising that the new immigrants end up in sectors demanding less highly skilled workers.

California high school graduation rates have been declining. The graduation rate for the native-born population has remained relatively constant at 97%, but the rate for the non-native-born population has dropped from 86% to 77% between 1980 and 1990 (U.S. Bureau of the Census, 1993). Twenty-six percent of California adults aged 20 to24 in 1990 lacked a high school diploma, the highest proportion in the nation and much higher than in the southern states, where high school graduation rates are traditionally low (Figure 3.8). This figure is a direct effect of large-scale influx of new migrants without a high school education in their own countries and of high dropout rates.

Does it matter that the rates are so high? Is this simply an adjustment that occurs whenever a new immigrant population arrives? Probably it does matter, if we recall the argument about the increased need for a skilled workforce. The high proportions of those who have not completed high school will increasingly bifurcate the workforce into a poorly trained and poorly paid group, on the one hand, and a highly educated, skilled, and well-paid group, on the other. The data comparing California with the United States overall suggest an increasing gap between California and the country as a whole (Figure 3.9). Members of the 20–24 age group are 10% less likely to be high school graduates than the U.S. population as a whole; not until age 45 does California show an educational "advantage."

Poverty

Immigrants not only begin childbearing in adolescence and fail to complete high school, they also have high levels of poverty. Because parental income is a critical factor for success in later life (Haveman and Wolfe, 1995), these high poverty levels have an impact beyond the immediate effects on the family members: they exert a long-term influence on the ability of individuals and families to create human capital. The data suggest that the poverty rate in California has increased as a direct result of large-scale immigration (Figure 3.10). New immigrants are more likely to be living in poverty than immigrants who entered the country earlier. Accordingly, the recency of the influx to California has increased the state's poverty rate by 19% (U.S. Bureau of the Census, 1993). Using data from tax returns, Thom (1995) found that between 1987 and 1991 tax returns for households' earning less than $20,000 increased by 11%, while the number of dependents in those households increased by 38% and accounted for 60%

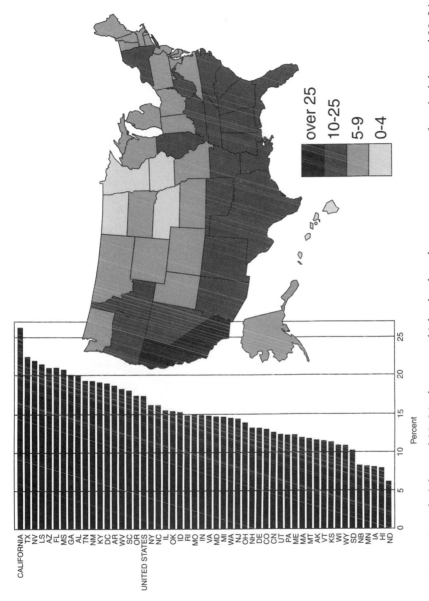

FIGURE 3.8. Proportion of adults aged 20–24 who are not high school graduates as a percentage of total adults aged 20–24. *Source:* U.S. Bureau of the Census, 1993.

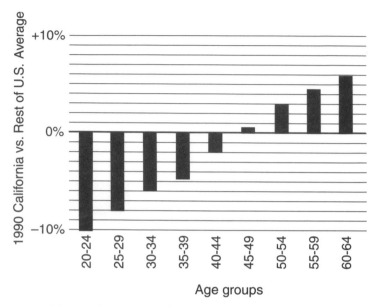

FIGURE 3.9.   The gap between California and U.S. high school graduates. *Source:*
U.S. Bureau of the Census, 1993.

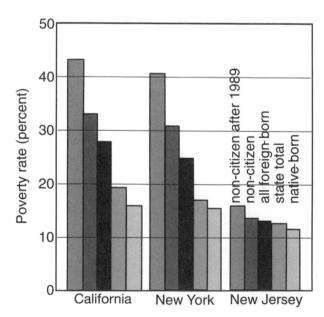

FIGURE 3.10.   Poverty rates by nativity and citizenship in 1990. *Source:* U.S. Bureau of the Census, 1990.

of all the added dependents for the state (California Franchise Board, 1992). In other words, the number of low-income and poor households increased significantly.

The critical long-term issue, however, is the change in child poverty. The proportion of children in poverty in native-born families in California increased from 12.3% in 1980 to 14.0% in 1990. Child poverty in immigrant families increased from 27% to 32.4%, and immigrant families accounted for 54% of the increase in child poverty between 1980 and 1990. Much of the rest of the increase was found among the citizen children of immigrants, mostly either Hispanics or Asians. Only 6% of the increase in the number of poor children occurred among whites and African Americans (U.S. Bureau of the Census, 1993).

At the same time, the levels of child poverty vary considerably across immigrant groups. Child poverty rates for immigrants from Japan, the Philippines, Korea, Europe, and Israel are all less than those for the state as a whole. Poverty rates for children in refugee Asian groups from Vietnam, Cambodia, and Laos and for children from Mexico and Central America are all above the average statewide rate. Almost half of the children in families of Mexican origin are living in poverty; the implications for low levels of future human capital for this group need not be restated. Immigrant families' poverty has increased, and the numbers requiring state aid have grown as a result.

Levels of dependency in California rose significantly between 1980 and 1990. Aid to Families with Dependent Children (AFDC) increased by more than a half-million persons; from 1979 to 1995 the increase was 1,045,151, approximately 177%. Most of this growth is related to immigrants' or to their citizen children. The greatest proportion of the increase is due to the large number of citizen children (with immigrant parents) in poverty. In contrast, AFDC to black households increased by less than 5%.

The state as a whole is affected directly by the surge in migrants, but the effects are not spread evenly over the counties in the state. Impacts in entry-point cities and their counties, especially Los Angeles and San Francisco County, are much more significant than in some of the suburban counties in these metropolitan areas. As I will make clear, however, no community is unaffected.

COUNTY AND COMMUNITY PATTERNS

Cities and communities have changed in response to the changing flows of population. Many cities are more diverse than the state as a whole, and 10 cities now have no single racial or ethnic majority (Figure 3.11). Other cities will follow this path as immigrant groups have children and add to

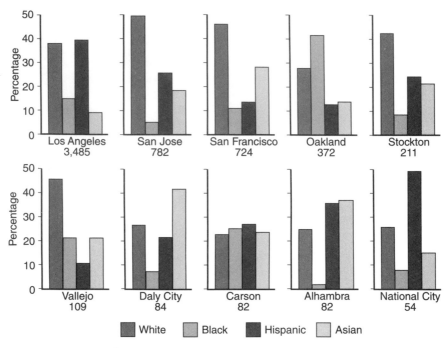

FIGURE 3.11. California cities without an ethnic/racial majority, 1990 (cities of >50,000 population in 1990). *Source:* U.S. Bureau of the Census, 1990.

the state's population. The entry-point cities, Los Angeles and San Francisco, are divided more sharply between white and Hispanic or white and Asian than some of the other cities, where the four major racial/ethnic groups are distributed almost evenly. In most cases, however, one group is numerically a majority. Only 20 years ago, these cities were mainly non-Hispanic white communities. Given the projected population increase due to Hispanic births, it is likely that many cities will become majority Hispanic in the next two decades.

Communities change as their metropolitan areas change. Communities and their residents age in association with one another, and a portion of the population changes each year as people move in and out. This is a measured process in which younger families replace members of the older population who die or move away. An influx of new foreign-born immigrants, who are usually younger than the population in the neighborhood, leads to a more rapidly changing ethnic composition in the younger age groups. Thus the influx of foreign-born migrants has changed the look and feel of Burbank, Glendale, Carson, Daly City, and Vallejo, making them quite different from how they were three decades ago.

Because the change occurs over decades and across generations, plots of the different ethnic groups by age for sample communities in Los Angeles and San Francisco show just how a community changes from the "bottom end" (Figures 3.12 and 3.13). The new young immigrants affect the younger age brackets first because the influx consists of mostly youthful new immigrants and their larger families. The native-born population or earlier waves of immigrants grow older; thus the graphs exhibit a concave effect toward the older ages.

In 1960, except for neighborhoods in south central Los Angeles, most communities in the Los Angeles metropolitan region were overwhelmingly Anglo. Only a few communities had any black or Hispanic presence. Even in 1970, the first time period shown on the graphs, most communities were still overwhelmingly nonminority. Exceptions included the black communities in Carson and the rapidly growing Hispanic communities in Monterey Park and Alhambra.

The change across the age structure is important because of the way in which change occurs in communities. It has implications for local politics and for the services that communities provide. Young Hispanics want more funds for schools, while older white and black households are concerned about healthcare and health services. The potential conflict is illustrated in Southgate: in 1990 the city was almost totally Hispanic, but it was still majority white for the age 65 and over population.

The picture is slightly different in Northern California, but the structural change is the same. In the communities in this part of the state, the shifts are less dramatic and involve more even change across the age groups. Even so, close to half of the younger age groups are members of minority groups. The growth of the Asian groups is notable, especially in Daly City and Milpitas (Figure 3.13). Daly City is 85% minority in the school-age population but 40 to 60% white in the over-60 population. The growing school-age population needs additional resources for many of the limited English proficiency (LEP) children, while the aging white population, who are the voters, are concerned about Social Security and entitlements for the elderly. The situation in these communities today is different from that at the turn of the century, when the elderly population of the United States was far smaller than it is now.

The graphs indicate that the immigrant centers have become much more diverse over the past three decades. They also show that the state's black population has diminished in relation to the population size of other groups and has also become older. In addition, they emphasize the decline and the aging of the non-Hispanic white population in cities in the Los Angeles metropolitan region. Alhambra, Glendale, and West Covina are still predominantly white "older" communities, but predominantly nonwhite "younger" communities. The graphs thus present an arithmetic picture of

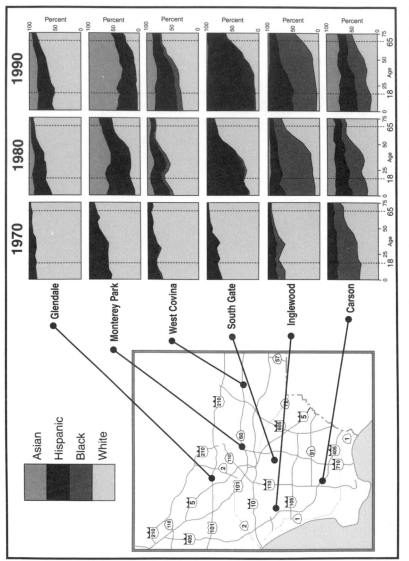

FIGURE 3.12. Community ethnic change over time for selected Los Angeles County cities. *Source:* U.S. Bureau of the Census, 1972, 1983, and 1991.

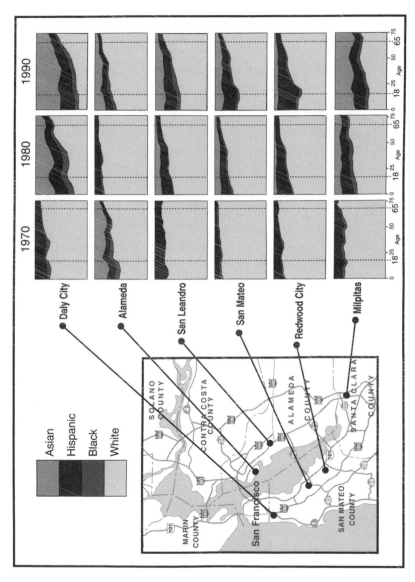

FIGURE 3.13. Community ethnic change for selected Bay Area Communities. *Source:* U.S. Bureau of the Census, 1972, 1983, and 1991.

the future as the younger age cohorts work their way through the system. The data have implications for community cohesiveness, community cooperation, and community resources, and for the process of assimilation.

IMPACTS ON THE COUNTIES

The selectivity of immigrant's destinations—the fact that different groups locate in different parts of the state—creates variations across counties and metropolitan areas. Some counties are affected more strongly than others, but all counties in California have borne, and continue to bear, the effects of sustained immigrant flows. These include the effects of elevated birth rates, changing healthcare needs, increases in the numbers of school-age children, and more families in poverty.

The changes in population and the effects of these changes are not uniform throughout the state. For example, Merced and Placer Counties each contained about 20,000 persons, but 20% of Merced's population was foreign-born and more than half of all the children born in Merced County between 1980 and 1990 were born into families with incomes below the poverty line. The localized effects at the county level are fairly obvious. More children in distress will require a greater proportion of county resources for schools and remedial help.

Births and Healthcare

The proportion of births to mothers who are foreign-born varies from a high of nearly 60% in Los Angeles County to only a few percent in small rural counties (Figure 3.14). The pattern of foreign-born births naturally follows the pattern of the foreign-born population. As we would expect from the large proportion of Mexican-origin women and their high fertility, a large proportion of all births occur to Mexican mothers. In 22 counties they account for more than three-quarters of the births. The agricultural counties are particularly notable for the concentration of Mexican-origin births. The corollaries, however—of healthcare and Medicaid-funded deliveries—mark the striking difference between Northern and Southern California. In 1993 the Medicaid-funded deliveries in the five northern counties hovered around 22% of births to foreign-born mothers. In the five southern counties, the average was nearer 40 % (Figure 3.15).

Births to foreign-born teen mothers exhibit no clear-cut regional patterns, though the rates are very high in the agricultural counties (Table 3.2). Such births account for almost half of all teenage births in Los Ange-

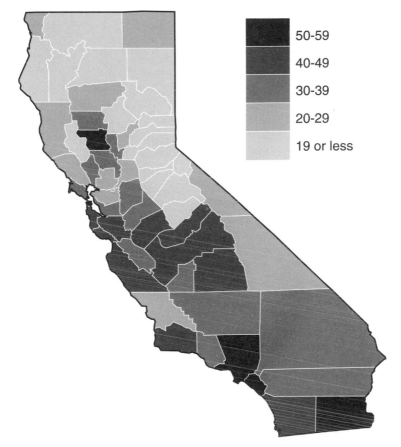

FIGURE 3.14. Percent of births to foreign-born mothers in 1993. *Source:* Department of Health, State of California, 1995.

les County. In Orange County, foreign-born mothers account for more than half of all teenage births; nearly half of these births were to Mexican-born teenagers. In several other counties (Santa Cruz, Santa Barbara, and San Mateo), Mexicans account for more than 40% of all teenage births. The teenage birth rates are now almost the same for Hispanics (16.1%) as for blacks (18%), and much higher than for whites (7.5%). Yet because there are so many more Hispanic than black teenagers, the number of Hispanic teenage births is more than five times as great (California Senate Office of Research, 1995).

High birth rates, large numbers of teenage births, and large numbers of Medicaid-funded births have both immediate and long-term effects. In the short run, these births increase the costs imposed on local healthcare

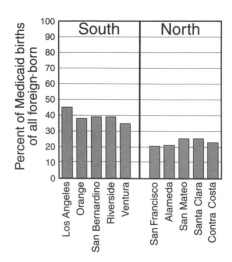

FIGURE 3.15.   Differences in Medicaid births for Northern and Southern California, 1992. *Source:* Department of Health Services, State of California, 1993.

TABLE 3.2. Teenage births in selected California counties in 1993

County	Total teen births	% of total births to mothers who are	
		teenagers	foreign-born teenagers
North Counties			
Alameda	21,911	9.6	24.2
Contra Costa	1,101	8.7	24.9
Santa Clara	2,358	8.7	38.2
San Mateo	766	7.4	55.9
San Francisco	671	7.4	37.4
South Counties			
Los Angeles	23,178	12.2	47.4
Orange	4,612	9.1	56.0
Riverside	3,444	13.7	28.0
San Bernadino	4,633	14.4	20.9
Ventura	1,271	10.4	34.5
Agricultural Counties			
Fresno	2,772	17.2	37.0
Kern	2,104	16.8	23.3
Sacramento	2,610	13.5	19.8
Tulare	1,353	18.9	34.4
State	70,091	12.0	37.4

Source: California Department of Health Sciences, State of California, 1997.

systems. In the long run, the social costs of lost schooling, of raising children in very-low-income and often poor households, reduce the acquisition of human capital and impair these children's ability to successfully enter the labor market. This is the ultimate cost. There is substantial evidence that children born to teenage mothers do much less well over their life course. The extent to which large numbers of teenage pregnancies are clustered in a few counties implies serious future problems of dependency and poverty.

Education and Limited English Proficiency (LEP)

As the population composition of counties changes, the school districts in those counties are affected. The impact is greater in some districts than in others, but major inner-city districts have seen fundamental changes from white or majority black to multiethnic districts of Hispanic and numerous Asian groups. The Los Angeles and San Diego school districts changed from majority white in the late 1960s to predominantly minority districts by the late 1980s, and to predominantly Hispanic districts by the mid-1990s. In Los Angeles the non-Hispanic white, black, and Asian populations have all declined, and the district is now 70% Hispanic. The changes are similar if less dramatic in the San Diego school district, which is now about 30% Hispanic, 30% non-Hispanic white, and 20% Asian. The relative decline in the black population, which is detectable in Los Angeles and San Diego, is especially noticeable in the Oakland school district. In all districts Asian and Hispanic student enrollment has increased; these changes are replicated in smaller school districts, especially in rural areas in Kern and Fresno Counties.

These changes alone create a challenge for the individual school districts. Much of this challenge involves the large numbers of young immigrants; many of whom do not speak English well, even if their skill levels are increasing rapidly. Learning for nonnative speakers is likely to proceed slowly, even with bilingual education programs. In view of the language barriers and the relative poverty, it is not surprising to find extreme differences in achievement levels between the native-born and the new immigrants. Despite evidence that some of the new immigrants are acquiring new skills and reaching high achievement levels (Rumbaut, 1997), which I explore in Chapter 5, an increasing number of inadequately prepared students are finding it difficult to move into mainstream educational programs.

In only a decade, the number of limited English proficiency (LEP) students in California schools has increased by 141%, from about a half million in 1985 to 1,262,982 in 1995 (Figure 3.16). In the 6-year period from 1989 to 1995, the increase was 70%. In some school districts—Los Ange-

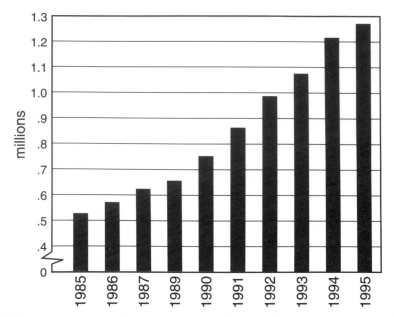

FIGURE 3.16.   Increase in limited English proficiency (LEP) students, 1985–1995. *Source:* Department of Education, State of California, 1995.

les, for example—the majority of the students are classified as LEP. Almost 24% of students statewide belong to this category; across counties the increase in LEP students has doubled and is more than 50% in almost every county (Figure 3.17). The most affected counties, with at least 100 LEP students each, are Kern, Madera, Lake, Marin, Napa, Placer, Sacramento, San Bernardino, and Sutter Counties.

The growing proportion of LEP students poses a fundamental challenge regarding the future demand for an educated workforce in California. The challenge involves a long-term dilemma: California workers must compete either by doing what workers in other countries cannot do, or by doing what those workers can do for lower wages. The latter is unlikely, given the large low-wage labor force in China and India.

Recent observers have suggested that to overcome the problems of a large LEP population, school districts will be required to "tailor testing" and to modify tests by removing the more difficult vocabulary items (Kenji Hakuta, quoted in the *San Francisco Examiner,* 15 May 1997). The role of bilingual education and the role of English are central questions in these discussions. Although educators are concerned with how to deal with disadvantaged children, the decision to deemphasize English at the same time that China and other Asian nations regard it as the lingua franca of global-

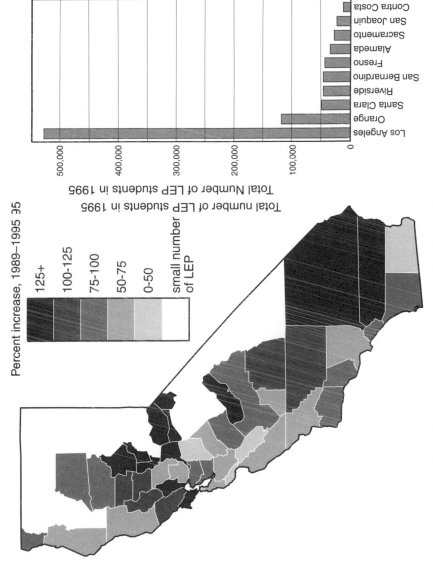

FIGURE 3.17. Change in limited English proficiency (LEP) students by California counties with more than 100 LEP students. *Source: of Education, State of California, 1995.*

ization is a matter of concern. Indeed, the growing gap between the wages paid to those with a high school education and those with a college education reflects the issue of LEP students in particular, and ultimately the issue of the future of state and county economies. The children of immigrants will become either productive, tax-paying adults or an enormous county and public expense. The obvious solution is to educate the students now rather than dealing with the problem in the future (Bean, Chapa, Berg, and Sowards, 1994).

In some counties the problem of LEP students has created tension with native-born African-American students, who may be disadvantaged although they are not LEP. We have almost no hard statistical data, and anecdotal reports are unreliable, but the graphs depicting community change show that many communities now contain substantial mixes of minorities, including native-born African Americans and new immigrants. These situations give rise to issues of resource use. In East Palo Alto in the San Francisco Bay area, the spending for bilingual students may be as high as $6,000 per student, while native-born students are funded at $3,900 per student. Yet there is still an achievement gap between native-born whites and black students. This situation is exacerbated in districts that have changed, in a decade, from 80% black to 70% Hispanic, as in several examples that we examined earlier.

The community change is expressed in yet another dimension, that of age. Demographic change has juxtaposed an aging nonminority community with a rapidly increasing minority population, even as the resources to educate a population with significant educational needs are dwindling at state and local levels. The aging nonminority population has been reluctant to pay the dollars necessary to improve the educational level of the state and its counties. The local effects are exacerbated where school districts are already experiencing cutbacks in funding, increased pupil–teacher ratios, decreased maintenance, and the suspension of language training.

Poverty and Dependency

Poverty and welfare dependency vary by county, but poverty among young children is a direct outcome of high levels of immigration and has the most dramatic implications for their future. Growing up in poverty severely reduces children's life chances. The poverty rates for young Hispanic and Asian children, who are overwhelmingly either immigrants or the children of immigrants, vary from 9% in Marin County for Asian and Pacific Island children to more than 50% in several Central Valley and Northern California agricultural counties, where there are large numbers of seasonal and often undocumented workers (Figure 3.18). Further evidence on the role of

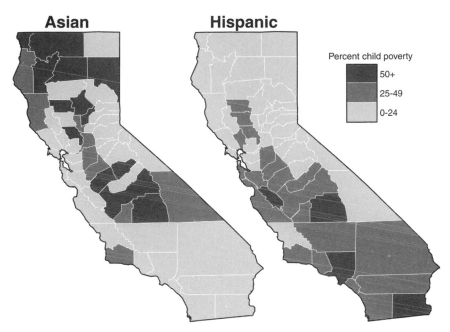

FIGURE 3.18.   Child poverty by county for Asians and Hispanics. *Source:* U.S. Bureau of the Census, 1993.

immigration can be gathered from the change in child poverty between 1980 and 1990. This change varies across California counties, but it is disproportionately Hispanic and Asian. Statewide, 94% of the increase involved Hispanic and Asian children; only 5% of the increase involved African-American children (Figure 3.19).

Welfare and assistance in general are contentious topics, and many observers have been concerned about showing that immigrants are not major users of public assistance. As we know, most of the total public assistance money spent in California goes to the native-born population. It is also true, however, that all immigrants now receive 44% of the total welfare funds spent in California (Table 3.3). Does the proportion vary by county? Where are the dependent populations?

The counties with the highest proportions of welfare-dependent populations are the Central Valley agricultural communities: Fresno, Sacramento, and San Joaquin, where 11.6% of the recent immigrant population received welfare in 1995. The proportion of welfare recipients in these counties is more than twice as large as in other Southern California counties. Los Angeles and San Diego are close to the state average for dependent recent immigrants; San Mateo is significantly below the average. Im-

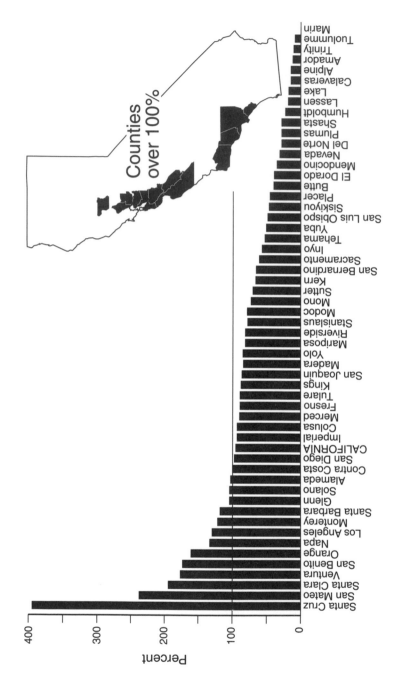

FIGURE 3.19. Hispanic and Asian proportion of increase in children in poverty (counties with more than 100% would have an absolute decrease in children in poverty without the increase of Hispanic/Asian poor children). *Source:* U.S. Bureau of the Census, 1993.

TABLE 3.3. Distribution of public assistance by ethnic background and ethnicity in California (in $ million)

Group	%	AFDC ($)	%	Total welfare ($)	%
Native-born					
White	47.8	971.8	24.7	1904.5	26.3
Black	5.5	524.2	13.4	864.6	11.9
Hispanic	17.5	794.7	20.2	1070.7	14.8
Asian	2.9	16.4	.4	125.9	2.3
Immigrant					
1980	10.1	1153.3	29.4	1988.8	27.5
1980–1994	15.4	406.7	10.4	1197.4	16.5

Source: Current Population Survey, 1994.

migrants from Mexico have the highest dependency ratios in Fresno and San Joaquin Counties, followed by their counterparts in the metropolitan areas of Los Angles City and Los Angeles County and in San Francisco. Overall, however, they receive only 10% of the total public assistance income in California, and they are not especially dependent in comparison with Asian and Vietnamese groups.

Despite the large numbers of immigrants living in poverty in metropolitan areas, rural California counties are also under considerable stress. Taylor, Martin, and Fix (1997) show that poverty is increasing in the towns of the Central Valley of California at the same time agriculture is experiencing increased prosperity. The low-paid migrant workers that once moved backward and forward from Mexico are now settling in or near the towns adjacent to California's rich farmlands in the Central Valley. The evidence seems to be that their wages are too low to help support local businesses and their public welfare needs are exhausting the local government services.

Asians, too, are a welfare-dependent population in the agricultural communities in the Central Valley, but they also make up large proportions of the welfare-dependent populations in San Francisco, San Diego (Vietnamese), and Sacramento. These groups also show very high positive standard deviations above the statewide dependency ratios, especially in the Central Valley counties. The greater welfare dependency of the Asian population in general is related both to the refugee status of some of the Asian population and to the fact that the family reunification process has brought in a large number of elderly family members. Clearly, the policies and outcomes regarding public assistance vary widely. Of course, the fact that recent immigrants are receiving public assistance income, and in substantial amounts, is noteworthy in its own right.

COSTS AND CONTRIBUTIONS

There is no more contentious issue than that of the costs and benefits of immigration. The debates about whether immigrants pay more in taxes than they receive in benefits have generated widely fluctuating net evaluations of the costs. Large-scale national studies by Huddle (1993) and the Urban Institute (Passel and Clark, 1994), and local studies in California by the General Accounting Office, have produced very different conclusions. The figures range from a net benefit of $2 billion dollars to a net cost of $18 billion dollars. Who is right?

On the one hand, let us consider an immigrant family of four, earning less than $10,000; some half-million of these families live in California. They may be paying $1,000 or $1,500 in taxes, but their two children in the public school system cost approximately $4,500 each, and much more if they are LEP students. The family also may be receiving some public assistance and may use local health facilities. Clearly, whatever taxes they are paying does not balance the cost of services they receive. On the other hand, it is difficult to measure the long-term economic benefits from the contributions of the immigrant population as they become members of an educated workforce and earn larger taxable incomes, which in a sense pay back the investment in human capital. In addition, they provide low-cost labor to the economy. Native-born low-income populations are in exactly the same situation, so this is not simply an immigrant issue.

The question of overall costs and benefits has no definitive answer. Local costs are not reimbursed by federal or state sources, though they are acutely apparent, and the stresses and inequalities they create are borne, often disproportionately, by local residents. A specific (though incomplete) analysis of these costs concluded that immigrants represent a net gain to the United States economy but create a deficit in Los Angeles County (Moreno, 1992). Immigrants and their children who arrived between 1980 and 1990 were calculated to generate about $139 million in revenues, taxes, and fees to the county and about $947 million in costs, for a net deficit of some $808 million. At the same time, the immigrants generated about $4.3 billion overall to the United States, with $1.5 billion in education costs; much of this amount was a state cost. This detailed example demonstrates why estimating net effects is so controversial.

California currently spends a little more than $5,000 educating each student, and the proportion of state revenues spent on public education is projected to rise substantially in the coming years. A major part of the increasing costs is related to the increase in undocumented children and in the citizen children of undocumented aliens. Estimating the costs of educating undocumented children depends on estimates of the number of undocumented aliens and the percentage of their school-age children enrolled in the public school system. Again we enter the realm of estimates and assumptions. Estimates of the cost of educating undocumented students in California ranges from $1.3 billion to $2.2 billion (Clark, Passel, Zimmerman, and Fix, 1994). The estimates by the State of California are considerably higher than those of the Urban Institute, and relate both to differences in estimates of the number of undocumented students and to estimates of the costs of per pupil education. The most plausible estimates suggest that 70% of all undocumented persons live in Southern California and that the costs of educating undocumented students alone may be as high as $1 billion. Yet even though the costs generally are borne by the state, the reimbursements do not cover all the costs of education. Thus local districts are affected to various degrees depending on their immigrant mix.

Local costs for healthcare are also rising. Although Medicaid is paid in part with federal dollars, there is increasing evidence that the local communities will be required to absorb a greater share of healthcare costs. Counties in California are legally bound to provide for destitute persons who are ineligible for any other public help, and must do so without discrimination respecting citizenship; therefore the counties may incur significant additional costs. Recent legislation has delayed the immediate effects of welfare reform, but according to estimates cited during the debates over reform, the state could incur additional costs approaching a billion dollars a year. For Los Angeles and Orange Counties, the costs would be about a half-billion dollars. The final effects of the ongoing changes are not yet evident, but it is clear that the costs will be redistributed unevenly.

SUMMARY

Aggregate changes have specific outcomes in particular communities. As a result of these immigration-induced spatial inequities, some communities bear disproportionate effects of aggregate migration while others escape the costs. These differential effects are greatest in the metropolitan centers of Southern California and in the Central Valley agricultural communities.[3] Most of the benefits are long term in nature, will come from enriched education, and may be reaped in destinations far from the places where the costs were borne. At the same time, we see increasing evidence of a growing division between immigrants who are "making it" and immigrants who are struggling at the bottom of the economic ladder. Without considerable economic and educational inputs, the latter will be unable to integrate into the larger U.S. society.

Perhaps the most important finding discussed here is that the continuing high fertility of the Hispanic population is likely to add a substantial number of children to low-income (if not poor) families. The very large increase in teenage births, in students with LEP, and in children requiring extra but unavailable health services may well overwhelm inner-city schools and public health services. Concentrated family poverty and dependency are increasing and are not likely to decline while the current birth patterns prevail.

NOTES

1. A map of counties, cities, and localities is included as an appendix.
2. Annual births in California have declined since the peak in 1990. The decline is associated with a falling birth rate and with the changing age structure of the population. The aging of the large younger cohorts in the California population, however, will lead to a reversal in the decline and to an overall increase in births beginning in about 2001 (Demographic Research Unit, Department of Finance, State of California, *Actual and Projected Births by County, 1970–2006*).
3. A recent book deals specifically with poverty in the agricultural communities of the Central Valley. Taylor, Martin and Fix (1997) show that extreme poverty exists even in communities that are doing well economically.

CHAPTER 4

IMMIGRANT EXPERIENCES IN CALIFORNIA

Whether the interviews and stories are published in the *Los Angeles Times,* the *Washington Post,* the *San Francisco Chronicle,* or the *Sacramento Bee,* they tell a similar tale, usually but not always involving Mexican or Central American migrants, who are struggling to escape the hand-to-mouth existence of low wages and poverty in their own country but who end up struggling in their new home as well. "People keep coming and it's hard to find work . . . it's very difficult . . . people are desperate" (*Washington Post,* 10 May 1997).

Are the homeownership success stories of immigrants who arrived before 1980 apparent in the labor market? Are recent immigrants earning as much as earlier immigrants? Are they able to assimilate into the U.S. labor market? We know already that the Immigration Reform and Control Act (IRCA) of 1986 legalized many low-income and poorly skilled migrants who had been living in the United States without documentation. We also know that after the legalization program, there was a surge in requests for family unification visas. Thus the poor and low skilled migrants brought in family members who were quite similar to themselves in income and skills. The overall poverty of these immigrants is less severe than the poverty of those they left behind in rural Central American and Southeast Asian villages, but without skills the future of these most recent migrants is uncertain. Their future is a central part of the debates over the immigration process.

The immigrant process is the subject of two debates. One debate is about whether recent waves of immigrants are faring less well than earlier waves, and whether, as a result, they are likely to fall into permanent poverty and dependency. The other debate focuses on whether the new immigrants are disadvantaging native-born low-wage workers. Both of these debates are central components of the discussions about our future policy on immigration and how many immigrants should enter the United States.

The initial studies of immigrants' adaptation to the U.S. labor market suggested that immigrants' earnings were lower in the years immediately after their arrival , but that in 10 to 15 years their earnings matched those of the native-born population (Chiswick, 1978). The research even suggested that immigrants not only caught up with the native-born but also, for the same skill levels, surpassed native-born workers earnings after a relatively short period in the United States. These findings, at first sight, are consistent with our ideas of immigrant arrivals in earlier periods. Those immigrants came with few skills, but through hard work and the application of the skills that they did possess, they were able to move up the earnings ladder. On-the-job training, skill certification, and learning English all played a role in creating more productive workers and, by extension, the potential for higher earnings and eventual prosperity.

Those who believed that immigrants actually surpassed native-born workers suggested that self-selection was at work. That is, the immigrants were "self-selected strivers"—the most ambitious, most energetic, most industrious individuals in their country of origin—and thus were likely to be more successful in their new locations (Simon, 1989). Their innate ability and drive, characteristics that are often cited in anecdotal reports today, explain why they were more successful than comparable native-born workers.

Unfortunately, more recent research has raised questions about the immigrants' "catch-up" behavior. A more careful examination employing longitudinal studies that tracked migrants over time, rather than cross-sectional studies that created a "snapshot" of migrants at a particular time, suggests that immigrants' gains were not as great as originally suggested and that an immigrant's lifetime earnings are unlikely to surpass those of the native-born population (Borjas, 1987). This does not mean that immigrants' children cannot catch up in skills, and thereby in earnings, but it does mean that their success depends on educational attainment.

The debates surrounding the analysis of earnings emphasize the fact that immigrant cohorts have changed substantially over the past 30 years. The figures in Chapters 1 and 2 demonstrated the significant changes in the origins of the population. The shift in origin is accompanied by a shift in skills and incomes. The debates also show that outcomes are likely to be quite different in different contexts. Studies in California for all immigrants reveal a decline in the earnings of recent waves of immigrants relative to the earnings of the native-born, although similar studies in New Jersey demonstrate that most immigrant groups have experienced both absolute and relative growth in earnings. That is, they were earning more in 1990 than at earlier times, and they were closing the gap between themselves and the native-born (Garvey, 1997). However, it should be noted that immigrants' outcomes obviously differ across regions, and that national studies do not capture the complexities of local outcomes.

In addition, immigrants are not a uniform group. If recent studies have accomplished nothing else, they have established that immigrants differ markedly in background, experiences, and outcomes. When immigrants are separated into groups, even into very broad groups, they follow very different paths. In this chapter I examine the changing pattern of skills, earnings, and dependency for recent immigrants as a whole and for Mexicans and Central Americans, Asians, and Middle Eastern immigrants as separate groups. Of course, the Asian and Middle Eastern groups contain subgroups with considerable differences, but the division I use here allows me to take some composition effects into account, and to provide detailed information on changes in immigrant groups over time.

THE PATHS OF NEW IMMIGRANTS

A substantial literature has attempted to disentangle the changes in immigrants' earnings over time (Borjas, 1994; National Research Council, 1997). The economic models that examine wage differentials between immigrants and the native-born population control for age, education, and gender, and consider whether immigrants are doing less well in these controlled situations. Such studies tell part of the story but provide much less information on the equally important question: How are immigrants doing as a group? As a result of changes in the *immigrant pool* are the new arrivals doing *as well* as earlier arrivals, or are they doing *less well*? In the aggregate approach employed here, I focus on the changes in the skills and earnings of the immigrant pool that entered the United States in successive stages since the late 1960s. The data for the analysis are drawn from the Public Use Microdata Samples (PUMS) from the U.S. Bureau of the Census for 1970, 1980, and 1990, and from the Current Population Survey for 1996.[1] For each of these years, I compare the measures for the native-born with those of recent immigrants—that is, for immigrants who arrived in the preceding 5-year period—between 1965 and 1970, between 1975 and 1980, between 1985 and 1990, and between 1990 and 1995. The years of education and earnings are computed for the total workforce aged 25–64. I also compute the fraction in poverty and on assistance for households in which the household head is at least 18 years old.

It is important to recognize both what I am attempting in this analysis and what I am not attempting. I am interested in average outcomes for new immigrants in comparison with earlier arrivals. In the first part of the analysis I am not controlling for age, composition, or skill levels to measure earnings. Indeed, immigrants with skills similar to those of well-trained native-born workers are more likely than less highly skilled arrivals to be earning a wage closer to that of the native-born population. I focus here on the aggregate outcomes in California as a whole, not on individual workers'outcomes. In the second part of the chapter, I use other data to study the pattern of success over time: I examine how immigrants are doing in particular niches in the job market in comparison with the native-born population.

The United States versus California

As reported by Borjas (1994) and Clark (1998a), the successive waves of immigrants to the United States are doing less well over time. Although these recent immigrants had gained almost a year in average years of education by 1990, they were still 1.3 years behind the native-born population. Recent immigrants earned about 16% less than the native-born in 1970, but more than 30% less in 1990 (Table 4.1). The poverty rate increased (in fact, while the native-born poverty rate declined, it nearly doubled for the immigrant population), as did the proportion of welfare-dependent households. In this context I examine the changes for recent immigrants in California; where possible I extend the analysis using data from the Current Population Survey for 1996.

For the analysis of the California data, I examine data for both men and women aged 25–64 and currently in the workforce. In California the native-born male population on average has 2 more years of education today than 20 years ago, and the proportion of native-born persons receiving public assistance has declined (Table 4.2). The poverty rate was stable until

TABLE 4.1. Skill and earnings levels for natives and immigrants in the United States

Group/variable	1970	1980	1990
Native-born			
Mean educational attainment (years)	11.5	12.7	13.2
Poverty rate (%)	13.7	12.2	12.4
Recent immigrants (<5 years in the U.S.)			
Mean educational attainment (years)	11.1	11.8	11.9
Differential from native-born	–0.4	–0.9	–1.3
% wage differential from native-born	–16.6	–27.6	–31.7
Poverty rate	18.8	28.1	29.8
% of households with public assistance	5.5	8.3	8.3

Source: Borjas, 1994, and author's calculations from PUMS tabulations, U.S. Bureau of the Census, 1972, 1983, 1992.

TABLE 4.2. Skill and earnings levels for immigrant males vs. native-born males

Group/variable	1970	1980	1990	1996
Native-born				
Mean educational attainment (years)	12.4	13.6	13.8	14.3
% below the poverty level	10.1	9.3	9.3	13.5
% of households with public assistance	3.4	3.1	2.7	2.4
All recent immigrants (<5 years in the U.S.)				
Mean educational attainment (years)	10.9	11.6	10.8	11.7
Differential from native-born	−1.5	−2.0	−3.0	−2.6
% wage differential from native-born	−40.3	−47.9	−55.5	−57.9
% below poverty level	24.0	29.5	30.7	28.1
% of households with public assistance	3.5	5.1	6.7	14.2
Recent Mexican and Central American immigrants (<5 years in the U.S.)				
Mean educational attainment (years)	6.9	7.5	8.0	7.9
Differential from native-born	−5.5	−6.1	−5.8	−6.4
% wage differential from native-born	−59.6	−62.8	−71.5	−72.0
% below poverty level	29.8	32.0	34.0	38.6
% of households with public assistance	9.8	2.0	1.1	4.4
Recent Asian immigrants (<5 years in the U.S.)				
Mean educational attainment (years)	14.3	14.0	13.1	12.8
Differential from native-born	1.9	.4	−0.7	−1.5
% wage differential from native-born	−42.7	−43.6	−45.0	−58.0
% below poverty level	31.6	28.7	26.5	25.0
% of households with public assistance	1.8	9.1	12.0	18.8
Recent Middle Eastern immigrants (<5 years in the U.S.)				
Mean educational attainment (years)	14.6	12.6	13.1	
Differential from native-born	2.2	−1.0	−0.7	
% wage differential from native-born	−48.8	−43.9	−43.2	
% below poverty level	33.3	33.1	32.7	
% of households with public assistance		6.5	15.0	

Source: U.S. Bureau of the Census, 1972, 1983, 1992, and CPS, 1996.

the most recent statistics were issued; these newer statistics almost certainly reflect the increasing number of poor citizen-born children of immigrants. For recent immigrants (those arriving in the successive 5-year intervals), mean educational attainment has fluctuated around 11 years and has increased recently, but continues to lag nearly 3 years behind the native-born population. Overall, the new waves of immigrants are in a less advantageous situation than earlier waves. If the difference in years of education has ceased to decline, as suggested by the 1996 results, it is stable at about an average of 3 years. The poverty rate has increased slightly, but the dependency rate has steadily increased and showed a major jump in the 1996 results. Finally, the results for California are strikingly different from the averages for the United States as a whole. The differences in California have become almost twice those of the nation overall.

An analysis of the total immigrant flows reveals how immigrants to California are faring as a group, but a full answer to the question of immigrants' success levels requires a disaggregation by place of origin. The data for successive waves of Mexican/Central American, Asian, and Middle Eastern male immigrants reveal quite different paths over time. No immigrant group matches the native-born population at the time of entry, nor would this be expected; moreover, we find striking differences among the groups and differences in the patterns of change over time (Table 4.2). Mexican and Central American immigrants are the most disadvantaged and have fallen the farthest behind in successive waves of movers (Figure 4.1). The small size of educational gains has kept these immigrants about 6 years behind the native-born population. The decline in relative wages for this group is substantial. They were earning about 60% less than the native born in 1970 but close to 72% less in 1995. Poverty has increased for Mexican and Central American immigrants, but their rate of welfare dependency is the lowest of the three major origin groups.

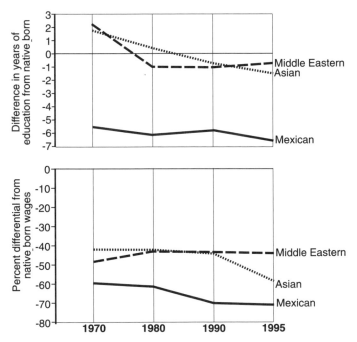

FIGURE 4.1.   Changes in skills and earnings for male immigrants. *Source:* Calculated from U.S. Bureau of the Census, 1992 and CPS, 1996 (see Table 4.2).

In contrast to the serious loss of relative position over time for Mexican and Central American immigrants, the Asian and Middle Eastern immigrants have fared much better. Asians' skill levels have declined somewhat, but their initial levels were higher than those of the native-born population. Actually, the difference in years of education between the native-born population and Asians and Middle East immigrants is quite small. The most recent relative decline for Asians, however, suggests that the very recent immigrants may be less well equipped than earlier waves to participate in the U.S. labor market. The effect of family reunification entries may account for this change. The decline in relative earnings is still modest for Asians, and in fact each wave of Middle Eastern immigrants improved its relative position (Figure 4.1). Thus the division of immigrant waves by origin is critical in understanding the immigrants' overall position. At the same time both the Mexican/Central American group and the Asian and Middle Eastern groups have high levels of poverty and increasing dependency (Table 4.2).

Women migrants also are doing less well than native-born women. The pattern of declining relative earnings over time applies to all women and to all the subgroups (Table 4.3). I report the data only from the PUMS because the sample data from the Current Population Survey are small. Immigrant women's overall earnings decline from about 34% less than native-born women to about 50% less. In general the declines for the subgroups are less than those for the equivalent groups of men entering in each successive wave (Figure 4.2). However, the Asian and Middle Eastern women immigrants who entered in the late 1990s earned much less than those who arrived in the late 1970s.

Even though women's relative earnings showed losses similar to the men's, immigrant women were less far behind their native-born counterparts than were immigrant men (Table 4.4). At the same time, the percentage of women in poverty and the number of welfare-dependent households headed by women were two to five times greater than for men. Of course, this information must be viewed in the perspective of men's and women's relative positions in the labor force. Native-born women on average in 1995 earned about 64% as much as native born men. Immigrant women in 1990 earned about 68% as much as immigrant men.

Again the story obtained by analyzing the earnings of successive waves of immigrants varies by immigrant group, by time of arrival, and by gender. The story reiterates the relatively disadvantaged status of Mexican and Central American immigrants and raises the obvious possibility that these waves of immigrants will find it very difficult to make any significant gains in the labor market. Asians in general did well, thought it is important to reiterate that the Asian migrants are divided further into the relatively well-educated Filipino, Korean, and Chinese immigrants and the

TABLE 4.3. Skill and earnings levels for immigrant women vs. native-born women

Group/variable	1970	1980	1990
Native-born			
Mean educational attainment (years)	12.4	13.6	13.8
% below the poverty level	10.1	9.3	9.3
% of households with public assistance	3.4	3.1	2.7
All recent immigrants (<5 years in the U.S.)			
Mean educational attainment (years)	9.9	10.8	10.4
Differential from native-born	–2.1	–2.2	–3.0
% wage differential from native-born	–34.7	–36.7	–49.8
% below poverty level	22.1	29.2	32.5
% of households with public assistance	12.9	12.6	13.9
Recent Mexican and Central American immigrants (<5 years in the U.S.)			
Mean educational attainment (years)	7.0	7.4	7.7
Differential from native-born	–5.0	–5.6	–5.7
% wage differential from native-born	–50.6	–52.2	–68.6
% below poverty level	27.4	34.7	39.8
% of households with public assistance	9.8	11.7	12.2
Recent Asian immigrants (<5 years in the U.S.)			
Mean educational attainment (years)	12.7	12.6	12.0
Differential from native-born	–0.7	–0.4	–1.4
% wage differential from native-born	–13.5	–29.1	–40.8
% below poverty level	22.2	25.5	23.9
% of households with public assistance	—	13.6	17.7
Recent Middle Eastern immigrants (<5 years in the U.S.)			
Mean educational attainment (years)	11.3	10.9	11.9
Differential from native-born	–0.7	–2.1	–1.5
% wage differential from native-born	+4.0	–30.9	–46.8
% below poverty level	33.3	27.5	29.7
% of households with public assistance	16.7	17.0	16.8

Source: U.S. Bureau of the Census, 1972, 1983, 1992.

less-well-established refugees from Cambodia, Vietnam, and Laos. The Southeast Asians, not the Hispanic migrants, make up the greater part of the dependent and poverty population. Borjas's (1994) finding of declining relative skills and lower earnings for immigrants to the United States as a whole is even more starkly evident in California for the very latest two periods of arrivals, but it is not the whole story.

If the immigrants entering in the late 1980s or the 1990s are generally doing so much less well than the native-born, what is the continuing pull for these immigrants? Clearly, many of the immigrants come as part of the family reunification program and are not motivated primarily by economic opportunities. At the same time, the evidence presented in Figures 4.1 and

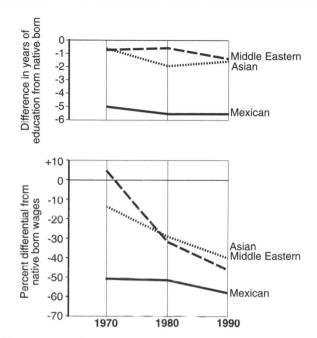

FIGURE 4.2.   Changes in skills and earnings for female immigrants. *Source:* Calculated from PUMS, 1996 (see Table 4.3).

4.2, especially for Hispanic migrants, attests to relative deprivation. The squeeze on incomes, however, is relative; when the analysis focuses on income adjusted for inflation, the picture of earnings effects for recent immigrants is quite different.

Immigrants are losing ground relatively, but only because the native-born population has done "relatively" well. Native-born males' earnings in California have increased (Figure 4.3). Of course, these figures are aver-

TABLE 4.4. Women's earnings as a ratio of men's earnings

	1970	1980	1990	1996
Native-born women/native-born men	49.6	52.5	60.3	63.5
White women/white men	48.8	50.7	58.1	60.7
Immigrant women/immigrant men	54.3	63.8	67.7	96.1*
Mexican women/Mexican men	60.6	67.5	66.4	60.0*
Asian women/Asian men	74.9	66.0	65.0	—
Middle Eastern women/M.E. men	100	64.7	56.4	—

*Samples are small and data should be interpreted with caution.
Source: U.S. Bureau of the Census, 1972, 1983, 1992.

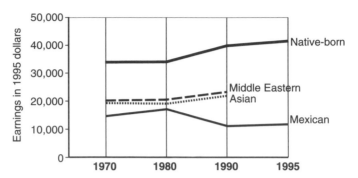

FIGURE 4.3.   Mean earnings in constant dollars for male immigrants and male native-borns. *Source:* U.S. Bureau of the Census, 1992.

ages, and not all groups have prospered in the changing California economy, but in general incomes were constant in the 1970s and have increased in the last 2 decades. A comparison of immigrants' incomes shows that in 1970 the new immigrants overall earned on average $20,520 in adjusted 1995 dollars. In 1995, immigrant men as a whole were earning about $18,000, a decline of slightly more than $2,000.This decline is much less dramatic than the relative loss reported in Figures 4.1 and 4.2; it represents a more modest 11.2% decrease in real dollars. The gains for native-born men, which increased by 21% from 1970 to 1995, caused the growth in inequality between immigrants and the native-born. Native-born women also earned more; they gained a little as measured against native-born men. Immigrant women who entered the United States between 1990 and 1995 had a slight gain in income (Table 4.5).

Other research on immigrants (Ortiz, 1996) provides similar results for Southern California. Despite modest progress in absolute terms, the data show continuing slippage in relative terms. Also, in keeping with the story told in this chapter, the progress was made only by those immigrants who arrived earlier and who entered the California economy in a more advantageous time.

Incomes reflect participation in the workforce, and participation, as measured by full-time employment, varies greatly (McCarthy and Vernez, 1997). In general, the participation rates for many immigrants vary from the mid-70% to the mid-80% range. Immigrants with the lowest earnings also show relatively low rates of participation in the labor force. Mexican migrants have a participation rate of about 70%, which is toward the low end of immigrant participation rates. Participation rates have decreased for recent Mexican and Vietnamese immigrants and increased modestly for recent European immigrants.

TABLE 4.5. Earnings in 1995 constant dollars, 1970–1995

	1970	1980	1990	1995
Native-born men	34,347	34,222	40,349	41,580
Native-born women	17,037	17,971	24,350	26,421
White men	35,760	35,722	42,551	44,961
White women	17,443	18,118	24,741	26,793
Immigrant men	20,520	17,832	17,944	18,217*
Immigrant women	11,323	11,377	12,233	12,600*
Mexican men	13,886	17,727	11,514	11,644*
Mexican women	8,421	8,590	7,647	6,983*
Asian men	19,694	19,301	22,185	—
Asian women	14,746	12,738	14,417	—
Middle Eastern men	17,557	19,199	22,921	—
Middle Eastern women	17,712	12,418	12,945	—

*Samples are small and data should be interpreted with caution.
Source: U.S. Bureau of the Census, 1972, 1983, 1992.

At least for Mexicans and Central Americans, earnings have declined.[2] Yet despite the general agreement about the decline in earnings, observers disagree about why earnings have decreased and welfare dependency has increased. Borjas (1994) points to increasing numbers of low-skilled workers. Sorenson and Enchautegui (1994), however, do not find declines in educational levels; indeed, the results I cited earlier in this chapter confirm the fact that absolute educational levels in general are increasing. Recent evidence points to the possibility of vanishing jobs in the niches that new immigrants have been filling. If low-skill jobs do indeed begin to "dry up," then the situation for arriving low-skill immigrants will become even more precarious. At the same time, these very low-skill jobs are dependent on the health of the economy. When California is doing well there are more "low end" jobs than when the economy is in recession.

In summary, today's immigrants are faring less well over time than earlier immigrants and especially less well than the native-born. Their skills are lower on average, and their earnings are declining, particularly in the case of Mexican and Central American immigrants. These figures, as well as the data on poverty and dependency, raise doubts about the prospects for success among these latest waves of immigrants. At the same time, the actual dollar incomes are much higher than the incomes that these immigrants can earn in Mexico, Guatemala, the Philippines, and especially the refugee-sending countries in Southeast Asia. This significant difference between the incomes that can be earned in developing countries and those available even to the lowest skilled and minimally trained immigrants is a powerful stimulus for continuing immigration. At the same time, the data on poverty and dependency, in combination with very low and declining relative earnings, tell a story of the future that contains all the elements of an impoverished underclass. The least skilled immigrants are better off by the standard of where they came from, but worse off than those they have joined, and they have reduced prospects for eventually realizing the traditional dream of prosperity and upward social mobility.

IMMIGRANTS' PROGRESS

The low earnings of recent immigrants relative to those of earlier arrivals raise a fundamental question: What is the future of these recent immigrants? To examine the potential trajectory, I have tried in these analyses to examine how the earlier cohorts of migrants fared. Are earlier waves of migrants catching up with the native-born population? Will the gap, even if it is widening over time, eventually close? This issue is fundamental for the future of migrants and the host economy.

The suggestion that immigrants eventually close the gap between themselves and the native-born population and earn the same amounts after a decade or two (Chiswick, 1986) would be encouraging. As I have already noted, however, this idea has been challenged. The evidence appears to support a less sanguine view of the earnings of immigrant cohorts as they progress through the labor market (Borjas, 1994; Lalonde and Topel, 1991). Later waves of immigrants apparently are not "catching up" in the same way that earlier waves of immigrants managed to achieve parity.

Evaluating immigrants' progress is not a straightforward process. To fully understand the extent to which immigrants achieve parity, one must follow a cohort over time. How are the immigrants who arrived between 1965 and 1970 doing in 1990 or 1995? How are the immigrants who arrived in 1975–1980 doing 10 years later? The results reported here for the immigrants' lifetime economic progress are drawn from a detailed cohort study of migrants distinguished by country of origin (Shoeni, McCarthy, and Vernez, 1996). The cohort method follows a particular age group who entered in a 5-year window as recorded in the microdata files of the U.S. Census. It gives a general picture of changes in immigrants' earnings, although it cannot deal with changes in the cohort due to out-migration either to another state or out of the country. Yet the available evidence does suggest that selective migration does not substantially bias the results.

In general, immigrants who arrived in earlier periods do better over time, although there are some exceptions and some groups do much better than others (Figure 4.4). Immigrants from Europe and from China, Japan, and Korea all fare much better over time: the gains for the earliest arrivals

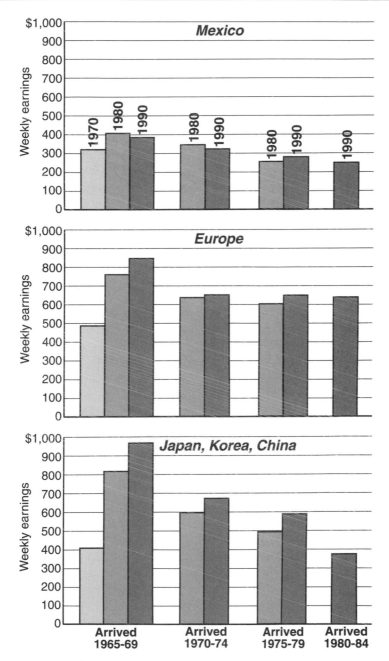

FIGURE 4.4.   Median weekly earnings of immigrants aged 25–34 by year of arrival for the years 1970, 1980, and 1990. Adapted from: Shoeni, McCarthy, and Vernez, 1996. Reprinted by permission.

are substantial over the 20-year period, as Figure 4.4 shows. European migrants who were 25 to 34 in 1965–1970 and who were 45 to 54 in 1990 registered gains of nearly 74%, from just under $500 per week to over $800 per week. Entrants from Japan, Korea, and China gained more than 100%. The gains are more modest for later arrivals from the source countries in Europe and Asia. The difference in the bars is much less over time; although the Europeans are holding their own, the new immigrants from Japan, Korea, and China are not. The story is quite different for both earlier and later arrivals from Mexico, who made small gains in the first decade, but who lost ground in the second decade. Mexican immigrants who arrived in the late 1970s made modest gains, but after a time they registered losses or only very negligible gains.

When the immigrants' earnings gains are compared with those of the native-born population, the experience of migrants from Europe and selected Asian countries is very different from that of migrants from Mexico. Immigrants from Europe and from Japan, Korea, and China who entered during the late 1960s increased their earnings significantly over time, as shown by positive percentages on the graph (Figure 4.5). In fact, the gains for early entrants from Europe and Asia were sufficient to bring their wages to parity with those of the native-born. Those from Japan, Korea, and China who arrived later, in 1970–1974, however, did not make such great gains and did not reach parity with the native-born. In fact, later European arrivals actually lost ground slightly. We find small gains for the Asian and European immigrants who arrived in the late 1970s. The changing rates of gains for groups of immigrants who generally bring greater skills and have more education may be an indicator of changing paths for migrants. Yet it is unlikely that the substantial earlier gains will be achieved as easily by later immigrant groups.

The paths for Mexican immigrants is very different. Mexican immigrants who were aged 25–34 and arrived between 1965 and 1970 earned about half as much as the native-born population and increased that percentage by only 2 points in the 10 years to 1980. In the next decade their earnings declined in relative terms to less than half those of the native-born. This was true for later entrants as well. As Mexican immigrants became older, they did not improve their position relative to native-born workers of the same age. More troubling, their relative position has fallen to about one-third that of native-born workers.

Do the results discussed above suggest that the rate of economic progress has changed? By comparing the relative positions of earlier and later arrivals—those who arrived between 1965 and 1970 and those who arrived from 1975 to 1980—one can estimate how well successive groups are doing. Both early and later arrivals from Europe and from Japan, Korea, and China have improved their position. As I noted above, however,

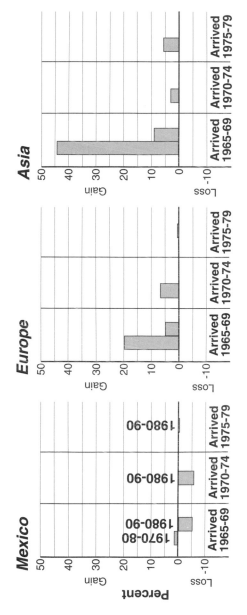

FIGURE 4.5. Percent change in the relative position of immigrants workers who were 25–34 in 1970, 34–44 in 1980, and 45–54 in 1990 against native-born workers of the same age. Adapted from: Shoeni, McCarthy, and Vernez, 1996. Reprinted by permission.

there are signs of a slight decrease in the rate of progress for these Asian immigrants. In contrast, the decline in Mexican's rate of progress is quite steep (Figure 4.6). While the arrivals from Europe and selected Asian countries tend to reach parity with the native-born population in 10 years or less, the evidence increasingly suggests that Mexican migrants are not even close to reaching parity in earnings.

Other evidence about changes in earnings over time confirms the decline in progress, especially for Mexican and Central American immigrants (Shoeni, McCarthy, and Vernez, 1996). According to profiles or pictures of the path of earnings for immigrants who arrived in the United States at age 25 in comparison with those for the native-born, we find (not surprisingly) that immigrants with more education are likely to make greater gains in earnings over time (McCarthy and Vernez, 1997).

The profiles follow three general patterns. Europeans and Canadians move above the native-born path by age 32 or 33, and, in the case of Canadians, continue to increase their earnings in relation to the native-born. Japanese, Koreans, and Chinese reach parity in their late 30s, but Mexicans begin well below the native-born and actually lose ground as they age. These profiles, which are based on earlier waves of migrants who were closer to parity in earnings when they embarked on their earnings path, are a useful indicator for the future pattern of immigrants' earnings. If the profiles are used as a base, the issue again is the future trajectory of a very large segment of the California population, a theme that also emerges in other contexts throughout this book. As Shoeni, McCarthy, and Vernez (1996) stated succinctly, a very large number of migrants have very low earnings, and their wages will not improve during their working lives. The labor market performance of a sizable segment of the California population may be inadequate to sustain their families. At the same time, we are

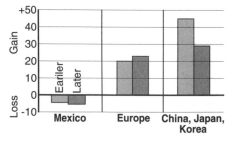

FIGURE 4.6. Changes in relative weekly earnings (relative to native-born) earlier and later arrivals of 25–34-year-olds. Adapted from: Shoeni, McCarthy, and Vernez, 1996. Reprinted by permission.

faced with the continuing paradox that it is this low-wage labor force that is providing significant inputs into the California agriculture, clothing, and electronics industries and that may be keeping California competitive with foreign producers.

In summary, the data show the consequences of differences in human capital, either in terms of what immigrants bring with them or of what they acquire in the United States. Thus the investments in human capital made by countries of origin influence the occupation and earnings trajectories of immigrants in the United States. The data on educational training in Mexico and El Salvador are very different from those for Japan and Korea (Table 4.6). In El Salvador and Mexico, 8% and 20%, respectively, of the children were receiving secondary education; in India, Japan, and Korea, these percentages vary from one-third to almost one-half. These differences are well known, but it is important to reiterate that the levels of training are important in explaining different earnings paths. However, education alone cannot explain the difference in success levels. Other personal characteristics of migrants (such as their English-language ability) and external characteristics (e.g., whether the economy is expanding or contracting when immigrants arrive) also play a role.

IMMIGRANT EFFECTS ON NATIVE-BORN WORKERS' WAGES

A second controversial issue concerns the impact of large-scale immigration on the wages of the native-born, especially on the low-wage native-born population. The recent consensus seems to be that the negative effects

TABLE 4.6. Number of students by country and level of schooling

	Primary	%	Secondary	%
El Salvador	1,031,559*	86%	92,858	8%
India**	95,740,000	63%	49,441,0000	33%
Japan	9,373,295	41%	10,992,498	49%
Korea	4,868,520	46%	4,559,557	44%
Mexico	14,401,588	69%	4,190,190	20%
Philippines	10,427,077	66%	4,033,597	26%

*This number includes children enrolled in preschool.
**Data from 1988.
Sources: Statistical Yearbook for Latin America and the Caribbean, 1996 Edition; 1996 Statistical Yearbook for Asia and the Pacific; 1994 Estadisticas Historicas de Mexico; 1991 Japan Statistical Yearbook; 1991 Korea Statistical Yearbook; 1994 Philippine Statistical Yearbook.

for the United States as a whole are small. Friedberg and Hunt (1995) conclude that a 10% increase in the fraction of immigrants in the population results in about a 1% reduction in the wages of the native-born. Lalonde and Topel (1991) argue that immigrants are likely to have stronger effects on each other than on the native-born. That is, the competition *among* immigrants is likely to be greater than the competition *between* immigrants and the native-born. They found that the greatest effect of recent migrants was to decrease the earnings of male migrants who had been in the United States for 5 years or less. Yet this effect was small: only about one-third of a percentage point decrease resulting from a 10% increase in recent migrants of similar status. The authors concluded that immigrants' impact on the labor market is insignificant. In addition, a study conducted recently in New Jersey, which suggests overall positive effects of large-scale immigration, found no evidence that immigration had negative effects on the relative wages and employment of the native-born working population (Butcher and Piehl, 1997).

In contrast to the above findings, the evidence from the National Research Council (1997) study of immigrant effects, recent work on California by McCarthy and Vernez (1977), and a study by Camarota (1998) point more directly to the negative impact of immigrants on the wages of native-born persons with low levels of education. Those studies and other work (Jaeger, 1994) suggest that about 40% of the decline in real wages of high school dropouts is attributable to the effect of immigration. Camarota (1998) shows that the effects are especially important for low-skilled native-born blacks and Hispanics. Moreover, although the wage gap closes somewhat for some groups, as I demonstrated in the previous section, the gap does not close at all for Mexican immigrants. Still, it appears that the effects are specific to location and to immigrant groups and are more likely to appear in earlier waves of immigrants than in the native-born. These results reiterate the richly nuanced effects of immigration and how it is played out locally in California and the United States.

The evidence from California on immigrants' occupational concentration and on changes in wage gaps increases our understanding of the way immigrants are performing and the associated affects on native-born residents. Immigrants are a sizable proportion of all employment sectors, and they have increased their penetration of all areas of employment (Figure 4.7). For example, in 1980, immigrants in California made up 62% of all workers in the garment industry; this proportion increased to 77% by 1990. In other words, immigrants dominate clothing manufacturing in California. But it is equally noteworthy that by 1990 immigrants accounted for at least 20% of every job sector except public administration and entertainment, and were close to 20% in the professional sector. The increases from 1980, by sector, ranged from 5% in public administration to

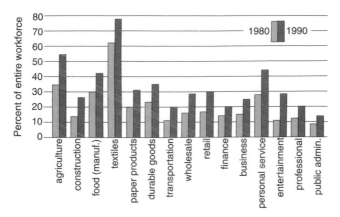

FIGURE 4.7.   Change in the immigrant proportion of employment sectors in California, 1980–1990. *Source:* U.S. Bureau of the Census, 1983, 1992.

almost 20% in agriculture. The most immigrant-intensive sectors, where immigrants were more than 40% of all the employed, were agriculture, food and textile manufacturing, and personal service. But in fact almost all sectors are now immigrant-intensive.

The associated changes in wages in the immigrant sectors have been examined in a variety of ways and with varying levels of sophistication. Most studies examine the wage gap between native-born white workers and immigrant and other ethnic workers, controlling for age and levels of education. Yet a more transparent graphical presentation of the wage gap (although without controls) tells the story equally well. A plot of the log wage gap[3] between native-born white males, on the one hand, and native-born white women, black men, black women, and Hispanic men and women, on the other, highlights those who have lost ground in the past 10 years (Figure 4.8). A positive value indicates that a particular group is making gains against native-born whites. A negative value indicates that a group has lost ground between 1980 and 1990. Those with the greatest relative losses are the native-born Hispanic males, precisely the group that is likely to be competing for jobs with new Hispanic immigrants. The wage gap was negative in all sectors for native-born Hispanics; in some sectors the gap was substantial. We find selected negative wage gaps in 10 of the 15 sectors for native-born black men but in only 6 of the 15 sectors for native-born black women (Figure 4.8). Hispanic women registered a positive gap in 8 sectors; that is, they did relatively better over the decade.

It is clear why it is so difficult to identify individual immigrant effects: immigrants are now a large proportion of the employment sectors. In addi-

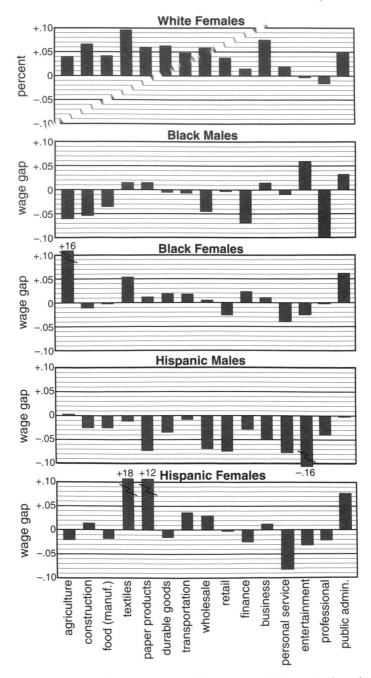

FIGURE 4.8.   Change in log wage gap with respect to white males for selected native-born populations, 1980–1990. *Source:* U.S. Bureau of the Census, 1983, 1992.

tion, the impacts vary widely by ethnicity, gender, and the occupational niche that particular immigrants occupy. The research shows that within employment sectors, certain groups of immigrants are even more highly concentrated (Waldinger, 1996; Wright and Ellis, 1996). At the same time, the overall wage gap across almost all segments of the Hispanic native-born male population suggests that the wages of this group are declining in the face of competition from new immigrants. This is true in many sectors but increasingly so in janitorial services, restaurant work, and construction.

The view that the wages of Hispanic native-born male workers are stagnant or declining is substantiated when we consider the change in weekly wages, in constant dollars, between 1980 and 1990. A plot of the change in weekly wages by gender, ethnicity, and employment sector reveals very different patterns across these groups (Figure 4.9). Hispanic males show declining or stagnant wages (defined as a change of less than 30 dollars a week) in 9 of the 15 employment sectors analyzed in this graph. Native-born black males' wages are declining in only 1 sector but are stagnant in another 5 sectors. Overall, where gains in weekly wages are realized, they are much smaller for native-born Hispanic and black males than for native-born Hispanic and black women (Figure 4.9). This descriptive analysis confirms other findings, which increasingly note the impact of immigrants on native-born workers, and complements my earlier analysis of the gains and losses among immigrants themselves.

REEXAMINING IMMIGRANTS' OUTCOMES

As a group, immigrants to California are losing ground in relative terms. Although some are holding their own in constant dollars, the relative and absolute decline for successive waves of Mexican and Central American migrants raises the very real issue of a bifurcated immigrant society and eventually of a bifurcated California society. The typical immigrant to California between 1985 and 1990 was Hispanic, had an almost 60% chance of earning less than $10,000 a year, only a 50/50 chance of working full time, and on average earned $2,000 less than the average migrant to the United States. The explanation for the increasing lag between the nativ born and immigrants in general and Mexican and Central American m grants in particular is a combination of low skills and the changing ec nomic contexts that I discussed in earlier chapters. Although Mexican an Central American migrants have lost ground relatively in the skills mea sures, they are not so different from earlier waves of Mexican migran who prospered. However, many of the well-paying low-skill jobs have var ished, for low-skilled immigrants as well as for low-skilled natives. A

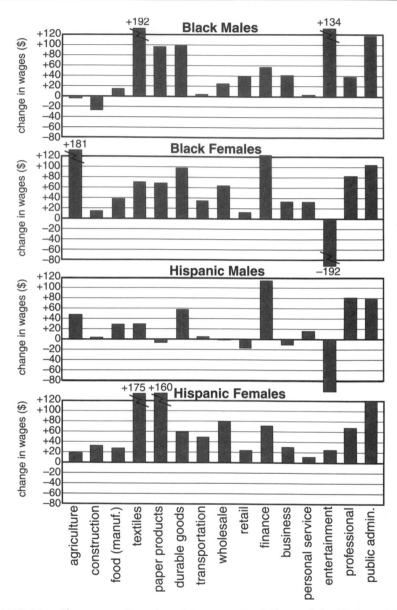

FIGURE 4.9.   Change in adjusted weekly wages (in dollars) of the black and Hispanic native-born employment sectors in California, 1980–1990. *Source:* U.S. Bureau of the Census, 1983, 1992.

though the economy expanded, the new immigrants from Mexico moved into the very lowest paying jobs in the expanding sector, jobs that hold little promise for better paying more stable employment, and eventual upward mobility. With their progress stalled, Mexicans and Central Americans cannot be optimistic in the same way that earlier immigrant waves of Mexicans were optimistic (Ortiz, 1996), or in the way that other immigrant groups appear to be optimistic about their future.

This chapter has emphasized the way in which barometers of opportunity are changing over time and across differing immigrant groups. Some recent migrant groups have done much better in the economic marketplace than other groups, though the distribution for none of the groups is even close to the overall income distribution for the United States as a whole. Europeans and Canadians, Koreans and Japanese, and migrants from the Caribbean and Africa are all surpassing other immigrant groups over time on measures of average income. Immigrants from Mexico, Central and South America, and Vietnam are lagging. And the impact issue comes up again because these latter, less successful immigrant groups are the groups that are especially concentrated in California.

It also appears that as a result of the lowered success levels for some groups that welfare dependency is increasing. While very few of immigrant arrivals before 1970 received public assistance, the broadening of entitlements and the increase in citizen children in immigrant households increased the number eligible for assistance. By 1995 nearly 15% of all immigrants who arrived (or were legalized by IRCA) received public assistance of some form. About 10% of Mexican immigrants who arrived in that period received public assistance. The public assistance dollars that went to immigrants went primarily to Southeast Asian households. Asian and Middle Eastern households had significantly increased their dependency by 1996.[4] Sixty-two percent of the general assistance dollars given to immigrants went to Vietnamese, Laotians, Cambodians, Thais, and Philippinos. Immigrants from China, Japan, and Korean received only 6.5% of the support for immigrants. Immigrants from Europe and Mexico received about 11% each. The emerging development of welfare support underscores a stark reality: lack of success in the labor market eventually translates into heightened dependency.

The paradox is that migrants are faring much worse even as the economy improves and generates employment. It appears likely that the growth in the economy is built on the dual pillars of the "high tech" information industry and the cheap labor of the sweatshops and agricultural corporations. Certainly, the influx of cheap labor is making it cheaper to hire homecare workers and gardeners and to purchase a whole range of goods and services much less expensively. But while this benefits the residents of affluent communities, such as the West Side of Los Angeles and St. Francis

Wood in San Francisco, the evidence suggest that the influx of low-income, low-wage immigrants is impacting the native-born low-skill population. The estimates of the impacts on the low-wage native-born tells conflicting stories, but Jaeger (1995) suggests that the impacts may be as large as one-third for specific groups. The economic future for new immigrants is clouded, and the notion that the market will take care of the problem is probably erroneous.

NOTES

1. The data on recent immigrants is derived from the U.S. Census 1970, 1980, and 1990 PUMS 5 % sample using the responses to the IMMIGR variable, "When did you come to the United States?," and selecting the codes for entry in the years 1985–1990 and 1975–1980. However, there is evidence that this may be overstating the true number of immigrants as some migrants already residing in the United States may have incorrectly noted a later arrival year. For example, Immigration and Reform Control Act (IRCA) immigrants may have answered yes to entry in 1988, the year of the amnesty program (Ellis and Wright, 1998). When the migration question from the census ("Did you live abroad in 1975–1985?") is used as a control on the immigration question, the numbers decrease to 3.5 million immigrants in the period 1985–1990 and 1.1 million immigrants in the period 1975–1980. Clearly there are issues related to undocumented migrants and circulatory migrants within the counting process. However, the exact effects on gross numbers have not been resolved, though Ellis and Wright (1998) suggests that past estimates of new migrants may have been severely overstated. As the data in this analysis are more directly concerned with variations across migrant origins and migration destinations within the United States, and as the results depend more on proportions than on raw totals, we will not explore this complication, except to note that there are some effects for particular groups by considering only migrants who said both, that they entered in 1985–1990 and reported that they lived abroad in 1985.
2. The results from this study are consistent with national studies by Borjas (1994), of Mexicans by Shoeni, McCarthy, and Vernez (1996), and of Hispanics by Sorenson and Enchautegui (1994). The latter reported that Hispanic immigrant men experienced a decline in real earnings in the 1980s.
3. The log of the wage gap, rather than the wage gap in actual dollars, is used to minimize the effects of the small numbers with very high wages.
4. Even allowing for errors from the small sample size of the Current Population Survey, the increase in dependency is notable.

CHAPTER 5

REALIZING THE AMERICAN DREAM

A 1990 *Time* magazine cover story on mid-21st-century America posed the question "What will the U.S. be like when whites are no longer the majority?" The answers already are partially apparent in California in the late 1990s. They range from the upbeat stories about immigrant entry to home-ownership and language acquisition to less positive accounts of low and marginal incomes for some recent groups of immigrants. Today's version of the American dream—owning your own home, securing a good education for your children, and putting down roots in the local community—has not changed for the new immigrants. How successfully are they at realizing this dream?

BECOMING A HOMEOWNER

Probably nothing is more central to the assimilation process than becoming a homeowner. Immigrants in the past started out as renters, but they or their children all aimed to own homes. Owning one's own home has long been enshrined at the heart of the American dream of independence and financial security, and is the surest badge of membership in the middle class. The dream has existed for the past half-century and has not been influenced substantially by rising house prices or stagnant incomes. For immigrants, as for native-born Americans, homeownership is equated with social status, the accumulation of wealth, and having "a piece of the pie." It is one of the important measures of "making it" in America.

There are sound reasons for pursuing ownership apart from issues of status, freedom, and flexibility. Ownership can furnish equity for financing businesses, loans for education, and wealth to pass on to children. These reasons apply as much to new immigrants as to the native-born popula-

tion. Although buying a house is an individual decision, it has many societal ramifications. Buying a house lowers the probability of moving, and
thus may be related to greater community and political participation. Yet,
because much of the new housing is located in suburban communities,
home buying has been linked closely to moving to the suburbs. Metropolitan areas spread rapidly after World War II as developers built new communities to accommodate the expanding baby-boom population. A large
proportion of that population grew up in single-family homes with increasing amounts of space, and developed a preference for owning such homes.
But buying a house is a relatively new phenomenon. As recently as the
1920s the United States was a nation of renters, not owners, but today
about 70 % of the households own rather than rent. It is only in the central
cities that ownership does not predominate.

Immigrants have the same ownership preferences as the native-born
population in California. Indeed, they are almost three times as likely as
native-born residents to cite home buying as their greatest priority (McArdle, 1997). This desire has not been dampened by high prices or low incomes. At the peak of the escalation in housing prices in California, surveys showed that more than 90% of the population under age 40 in Los
Angeles County wanted to buy a home, and that more than 75% of those
who wanted to buy expected that they would be able to do so eventually
(Heskin, 1983). Sixty-one percent of English-speaking Latinos and 81.5%
of Spanish-speaking Latinos wanted to buy a house in the next 3 years.
Even in Southern California, one of the tightest and most expensive housing markets, the push to own is strong, an indication that homeowning is a
central part of the transition in the new society. In the first section of this
chapter I examine who has been successful in making this transition to
ownership.

An Overview of Homeownership among Immigrants

In the 1990 Census California had 10.3 million households with heads
who were age 15 or older; 2.3 million of these were immigrant households.
More than 1 million immigrant families owned their own homes in 1990,
but the rates of homeownership vary a great deal by immigrant origin and
time of arrival.

It is useful to view California immigrant ownership rates in comparison to ownership rates among all households. Non-Hispanic whites have
the highest homeownership rate, 62%, followed by Asians, Hispanics, and
blacks (Table 5.1). These rates have been relatively stable over time. Black
ownership rates, however, have now declined even below those for Mexican immigrants who arrived before 1980.

TABLE 5.1. Percentage of households owning their own homes, 1980 and 1990

	1980	1990
Native-born	.58	.59
Ethnic group		
White non-Hispanic	.60	.62
Asian*	.53	.55
Hispanic	.42	.40
Black	.40	.36
All immigrants	.48	.45
Immigrants arrived before 1980		
Mexican/Central America	.35	.44
Asian*	.52	.59
Middle East	.43	.61

*Asian includes other races, not of Hispanic origin.
Source: U.S. Bureau of the Census, 1983, 1992.

The average ownership rate for all immigrants is 45% as of 1990 (Table 5.1). Thus almost half of all immigrants to California have managed to buy or are buying their own home. Nonetheless, immigrants' ownership rates have declined slightly in the last decade, almost certainly because of the very large influx of new immigrants who have not yet had time to save enough to enter the homeowner market. Mexican and Central American immigrants have the lowest homeownership rates among immigrants who arrived before 1980, but no group in the state has ownership rates as low as the black population.

Immigrant ownership rates vary widely by the immigrants' place of origin.
Immigrants from Western Europe and China have the highest ownership rates, almost two-thirds of these immigrants are homeowners (Figure 5.1). The lowest homeownership rates are those for Central Americans and Vietnamese. The data presented in earlier chapters, which demonstrated the large differences in income across immigrant groups, are the critical factor in immigrants' ability to buy their own homes and are a primary explanation for these differences in ownership. The Japanese, with their relatively high incomes, but low ownership rates, are an anomaly. To some extent, the ownership gap in that group may be related to the presence of students and the large number of employees of Japanese companies on temporary assignment in California. Even so, the fact that a large percentage of some immigrant groups have bought or are buying their own homes is a measure of immigrants' success in penetrating the real estate market.

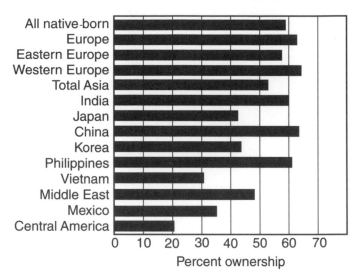

FIGURE 5.1. Ownership rates by immigrant origins, 1990. *Source:* U.S. Bureau of the Census, 1992.

Ownership by immigrants increases over time.

By 1990, ownership rates among immigrants who arrived before 1980 had increased significantly (Table 5.1). Immigrants from the Middle East who arrived before 1980 had a rate of .43; this rate had increased by 18 points in a decade. The gains for Mexican and Central Americans were smaller but still substantial.

At the same time, as in the native-born population, ownership varies significantly by age (Figure 5.2). Very few households with heads under age 25 are in the owner market, and only 20% of those between 25 and 30 own their own homes. The rate of ownership increases with age until about age 65, after which it declines.

Immigrants are not living in poor-quality housing.

More than 98% of California's housing units have complete plumbing and complete kitchens. Because the state's housing stock is relatively new, immigrants generally are not in old housing (U.S. Bureau of the Census, 1993). Only 11% of the houses occupied by immigrants were built before 1940. Even so, the housing they occupy is often overcrowded. Myers (1995; Myers and Wolch, 1994) points out that crowding has returned to Southern California as garages are converted illegally to living spaces and residents double up in existing apartments. In particular, Hispanic immigrants have became much more overcrowded in the last two decades. But while Mexican and Central American immigrants are overcrowded, levels

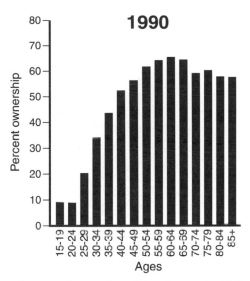

FIGURE 5.2. Ownership rates by age for all immigrants in 1990. *Source:* U.S. Bureau of the Census, 1992.

of crowding for other groups are quite low (Figure 5.3). Overcrowding (which we can also call "room stress"), a function of insufficient space in the dwelling, is often measured by the presence of more than 1.01 persons per room. On this measure, many immigrant groups lack sufficient space. Immigrants as a group have a index of 1.2 persons per room. To put the measure in context, the index for non-Hispanic whites for the state as a whole is approximately .7. Mexican and Central American owners have a rate of 1.7 and 1.8; the Southeast Asians, largely refugees, also show high rates of crowding. The lowest rates—that is, the greatest amount of space per person—are found among Europeans (Figure 5.3).

Such crowding is related in part to the relatively high cost of housing in California. The median housing price in California is more than $195,000, significantly higher than in New Jersey, for example, or for states in the Midwest. Because housing is so expensive, we would expect relatively greater housing-cost burdens in California, but this burden also varies a great deal by immigrants' place of origin. Again, the results indicate stress for Mexican and Central American households. The housing-cost burden can be measured as the percentage of income that is spent on housing payments; this proportion is about 9% greater for renters than for owners (Figure 5.4). The lowest cost burden among owners is that for Europeans, the highest is that for Koreans and Middle Eastern immigrants. Korean and Middle Eastern immigrants who are renters also have very

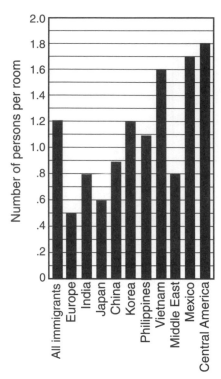

FIGURE 5.3.   Overcrowding among immigrant households, by group. *Source:* U.S. Bureau of the Census, 1992.

high housing costs relative to income: on average they allocate two-fifths of their income to housing. Vietnamese renters have a similarly high housing-cost burden.

The high housing-cost burdens among both relatively poor groups (Vietnamese and Central American immigrants) and relatively affluent groups (Korean and Middle Eastern immigrants) involve two different scenarios. On the one hand, Middle Eastern and Korean immigrants may be buying "more house" and thus investing a larger portion of their incomes, or they may be renting larger space to accommodate large families. On the other hand, Vietnamese and Central American renters, who have low incomes, are forced to spend large sums on rental housing and still may not be able to acquire sufficient housing space. It is even plausible to argue that two processes are working together: one process is acculturation to the American pattern of homeownership and suburbanization, and the other is the inability to move up the socioeconomic ladder.

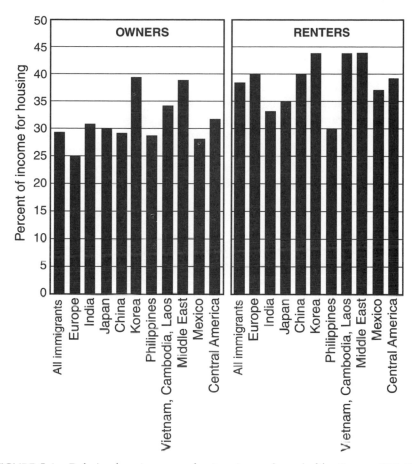

FIGURE 5.4. Relative housing costs for immigrant households. *Source:* U.S. Bureau of the Census, 1983, 1992.

TRAJECTORIES OF HOMEOWNERSHIP

Becoming a homeowner is tied closely to age and to income. In general, incomes increase with age. The rates at which individuals enter the homeowner market, however, varies considerably. To examine this process I look at ownership trajectories for all immigrants in comparison with the native-born in the United States as a whole, and then at specific patterns for Mexican/Central American, Asian, and Middle Eastern migrants.

Because changing tenure—moving from being a renter to being an owner—is closely related to income and resources, those immigrant groups who enter their new country with more resources, or who can translate their human capital into better jobs and thus more income, are likely to enter the homeowner market more quickly. Those with fewer resources and with more limited human capital will enter the housing market more slowly and will make slower progress over time. In addition, and in keeping with other findings reported in this book, earlier entrants are doing better than later entrants. For all immigrants who entered before 1970, 66% were owners as of 1990 (Table 5.2). In contrast, among all immigrants who entered between 1980 and 1990, only 22% were owners in 1990. Earlier immigrants from India reached overall ownership rates of 80%, and many other migrant groups were in the 70% range (Table 5.2). Conversely, the rate of ownership for recent immigrants was extremely low for Mexicans and Central Americans. These are the immigrants who are younger and have the lowest wages and fewest opportunities.

Overall, however, Table 5.2 shows that with time, immigrants, or at least immigrants who arrived by 1980, are achieving ownership rates similar to those of the native-born.[1] The homeownership rates in California are close to those for the country as a whole (U.S. Dept. of Commerce, 1996), and greater than those for New Jersey (McArdle, 1996). The very high levels of ownership even in an expensive housing market reemphasize the strength of the dream of homeownership across all ethnic groups. Ownership increases with duration of residence and reflects immigrants' overall

TABLE 5.2. Ownership rate by period of immigration

Place of Birth	Year of entry		
	<1970	1970–1980	1980–1990
All	.66	.45	.22
Mexico	.60	.33	.09
Central America	.51	.23	.08
Asia	.75	.64	.33
India	.80	.78	.37
Japan	.61	.53	.16
China	.79	.69	.44
Korea	.67	.56	.28
Philippines	.79	.67	.39
Vietnam/Cambodia	.60	.53	.18
Middle East	.68	.57	.30
Europe	.70	.57	.32
Eastern Europe	.70	.55	.27
Western Europe	.70	.58	.35

Source: U.S. Bureau of the Census, 1992.

belief that owning a home is a critical aspect of "making it" in the United States.

The rate and speed of the transition to homeownership are portrayed most clearly in cohort- trajectory diagrams. These diagrams show the rate of ownership for a particular group of similarly aged people as they grow older over time. Thus, if 50 out of 250 households who are headed by a person aged 20 to 24 are owners, the ownership rate is 0.2, or 20%. We can examine this same group/cohort 10 years later and determine how many are now owners. For example, the cohort that is now 30 to 34 years old has 100 owners out of 250, an ownership rate of 0.4 or 40%. Of course the cohort changes over time: some immigrants arrive, others leave, and some members of the cohort die. Yet these changes are relatively small relative to the size of the entire cohort, and the change in ownership over time gives a very clear picture of the process of moving into ownership with age. In the analysis that follows, I examine the paths for those who arrived before 1980 and discuss how that group as a whole progressed in the next 10 years. The technique has been discussed in detail by Pickles (1990), Myers and Wolch (1994), and Clark and Dieleman (1996).

Homeownership trajectories for the U.S. population as a whole are steep, and they plateau early; by ages 35–39, the population has reached the average ownership rate for the country. At ages 25–29, almost 45% of householders are owners; by ages 35–39, more than 65% have become owners. By the time household heads are in their late 40s and 50s, the ownership rate reaches a plateau in the 80% range (Figure 5.5).

For the native-born in California, the trajectories are less steep and never reach the rate or the speed of the trajectories for the United States as a whole (Figure 5.5). The ownership rates approach but do not reach the U.S. average of 80%. This difference is almost certainly due to two factors: the higher cost of housing in California and the large number of poorer native-born Hispanics who do not achieve the same ownership rates as the native-born non-Hispanic white population.

Trajectories of ownership for all immigrants as a group and for subgroups of Mexican and Central American, Asian, and Middle Eastern immigrants tell quite different stories, depending on the group (Figure 5.6). To reiterate: this section of my study *excludes* new immigrants to keep the data consistent; the trajectories are plotted only for immigrants who had arrived *by 1980*. Obviously, the trajectories for all immigrants average out the steep trajectories and high maximum levels of ownership for Asian and Middle Eastern homeowners and the much slower trajectories of the Mexican and Central American owners. Still, immigrants reach ownership rates comparable to those of the native-born, though they do so at later ages. They are in their late 40s and 50s before they make the jump to ownership.

The patterns for the three subgroups are very different. Mexican im-

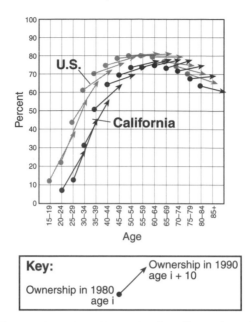

FIGURE 5.5. Cohort trajectories 1980–1990 for all U.S. households and native-born Californian households. *Source:* U.S. Bureau of the Census, 1983, 1992.

migrants' initial trajectories are quite slow because they entered ownership slowly. Even so, their rate of ownership almost doubled from quite a low base. However, they never rise much above 60% ownership even during the peak ownership years.

In contrast, Middle Eastern immigrants' rates of entry to buying a home are the steepest. One out of six households headed by someone aged 25–29 owns their own home; 10 years later, 6 out of 10 of these households are owners. In a 10-year period the cohort has increased its rate of ownership more than threefold.

Asian households in the cohort age 25–29 also increased their ownership. Asian immigrants who arrived before 1980 reached almost 80% ownership when they were age 45 or more.

Both Asian and Middle Eastern immigrants, however, display noticeable anomalies in their ownership patterns. Asian homeowners give up ownership after age 60, in a complete contradiction of the patterns of immigrants as a group. Anecdotal evidence suggests that this reflects a cultural pattern of moving in with their children. For Middle Eastern households it is possible that the homeowner market contains two different groups and two different paths. The younger Middle Eastern immigrants are entering

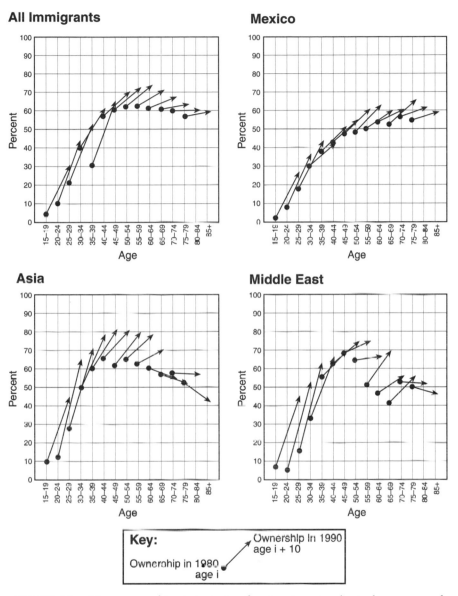

FIGURE 5.6. Homeownership trajectories for immigrant cohorts by age as of 1980. *Source:* U.S. Bureau of the Census, 1983, 1992.

the housing market aggressively, but the earlier immigrants, who are now in their 50s and 60s, are catching up only belatedly. Some are now in the elderly category and are moving away from ownership. The patterns reveal underlying cultural and economic differences, but the extensiveness of homeownership indicates the importance, to immigrants, of the move from renting to owning.

These trajectories for Californian immigrants overall are similar to results reported by Myers and Lee (1997) for Southern California but are quite different from the trajectories for immigrants in New Jersey (McArdle, 1997). In New Jersey the patterns for immigrants are more similar to the paths of the native-born population, and the immigrant groups appear to vary less widely. Myers and Lee (1997) use a more sophisticated technique than McArdle, but show a similar degree of advancement in Southern California, especially for Asians and whites. They also draw attention to the year of arrival as an important variable in the process of becoming a homeowner. Earlier arrivals, especially Hispanics, do better. Even though their starting point is lower, they advance rapidly in their housing careers. (See the data in Table 5.2.) Analyzing California as a whole rather than focusing only on Southern California reemphasizes the differences between immigrants' places of origin. Even though some Hispanics are doing well, many immigrants of Mexican origin are making only slow progress in the housing market.

The less steep trajectories for Mexican and Central American immigrants are certainly related to lower incomes and fewer assets. The steeper trajectories of many Asians and Middle Easterners may reflect a value system in which ownership is a higher priority. Mexican immigrants who arrived between 1975 and 1980 had mean annual incomes of $5,491 dollars, immigrants from China and Korea had average incomes of about $8,500, and Middle Eastern and European migrants had incomes of more than $12,000. That Asians' incomes are half again as large as those of Mexicans and Central Americans helps to explain the greater homeownership by Asian households. Asian households also may pool resources to buy into the homeowner market.

Asians who have lived in Southern California for some time have moved away from Chinatown and Koreatown to more expensive suburban areas. Abelmann and Lie (1995) report that poverty, not an identification with other Koreans, keeps Korean Americans in Koreatown. Korean Americans are a classic example of the movement of generally affluent populations to suburban locations in pursuit of "better" housing and higher quality schools. The move to the suburbs and to homeownership mirrors the earlier dream of white midwesterners for spacious, detached homes with room for pools and backyard barbecues. The dream of 1950s suburban California is pursued now by new immigrants; and when they can afford it, they move out of the old ethnic enclaves in the center of the city to new ethnic enclaves in the suburbs.

Korean Americans' residential moves are merely one aspect of a desire by all groups to move up and out, to escape immigrant clusters, which are often associated with high crime and poor environments for children. However, before we examine the pattern of upward and outward mobility, we must ask who enters the homeowner market and what roles age, income, and period of entry play in the ownership process.

THE DETERMINANTS OF HOMEOWNERSHIP

The process of becoming a homeowner is related to age, income, and the process of moving through the life cycle, that is, from living alone to marriage (or cohabiting). The birth of children increases the demand for more space and creates a tendency to move (as income increases) from less spacious rental housing to more spacious owner housing (Clark and Dieleman, 1996; Henderson and Ioannides, 1985; Olsen, 1987). The tendency to become an owner is not due to age per se, but rather to the "life course" events that are part of the aging process that leads to ownership. The process is the same for immigrants as for the native-born. The difference for immigrants in the United States, as opposed to immigrants in their countries of origin, is that hard work generally enables them to quickly approach native-born rates of ownership. Korean Americans, for example, "pride themselves on their large houses, unattainable had they stayed in Seoul" (Abelmann and Lie, 1995, p. 106).

Socioeconomic and demographic variables play the same role for immigrants as they do for the native-born. Age, income, and marital status are all important factors in becoming a homeowner. Being married increases the chances of homeownership regardless of citizenship status, but citizenship is an important part of the transition to becoming an owner (Callis, 1997).

To examine the relative roles of the forces that affect the move to ownership, I use a simple regression model of the probability of being an owner or a renter as a function of age, income, marital status, citizenship, and measures of the period of arrival (Table 5.3). The model explains quite well the likelihood of being an owner, fits well for each group of immigrants, and correctly classifies more that 80% of the cases. The coefficients in the table are standardized so that larger values indicate more important coefficients.

Income is the most important variable for all immigrants and for Asian, Middle Eastern, and European migrants, but not for Mexican and Central American immigrants. For Latino migrants, having migrated in earlier decades is the most important criterion for homeownership: the ear-

TABLE 5.3. Models of ownership

	All	Mexico	Asia	Middle East	Europe
Age	1.049	.906	.266	.531	1.204
Income	2.860	1.936	3.145	2.563	2.566
Citizen	.572	.179	.776	.570	.192
Married	1.030	.944	.953	.987	1.340
Immigration year					
'80–'90	−1.885	−2.801	−1.909	−1.540	−1.382
'70–'80	−0.857	−1.261	−.864	−.553	−.455
'60–'70	−0. 297	−.416	−.309	−.322	−.238
Gamma	.662	.646	.662	.594	.590

lier the arrival, the greater the likelihood of ownership. Later arrivals are much less likely to be owners. The coefficient is large and negative for migrants who entered in the 1970s or 1980s. In essence, only the earlier arrivals from Mexico and Central America have been able to enter the homeowner market.

We find a noticeable contrast between the coefficients for Mexico and for Asia. For migrants from Asia, the income effect is 30% stronger than it is for Mexican and Central American migrants (Table 5.3). Being a citizen is an important effect, the most important of any group. Age is important for almost all immigrants, but is not important for Asian buyers. That is, both younger and older Asian immigrants are entering the homeowner market. This finding for Asians may have been affected somewhat by the amount of foreign funds being shifted to the California real estate market. Recent migrants are less likely to be owners across all Asian groups, but this result may be influenced by the lower probability of buying by the less advantaged Asian groups, particularly Cambodians and Laotians. The very recent immigrants from the Middle East and Europe are more likely than other groups to be able to buy (as shown by smaller negative values of the "immigrated 1980–1990" variable). Citizenship is least important for Europeans and Middle Eastern immigrants.

SUBURBANIZATION AND HOMEOWNERSHIP

In my discussion of the house-buying behavior of Korean immigrants, I stressed the general desire to move from downtown to more affluent suburban communities. This pattern is common across all immigrant groups and in both Northern and Southern California. Studies of Hispanics' moves to suburban locations in the 1980s showed that those who moved out were also more likely to become homeowners (Clark and Mueller, 1988). Those earlier studies established that for Southern California, ownership increased with movement to suburban counties. At the same time, professional occupations were more prevalent and socioeconomic status was higher among those who moved to the suburbs.

Are those who move from the inner city—from Los Angeles and San Francisco—also improving their housing status? The ideal way of examining whether upward and outward mobility are related would be to use time-series data to trace second and third generations of immigrant families as they moved from cities to suburbs and to track their associated home-buying behavior. That information is available only with individual surveys, but as a proxy for measures of changing mobility and housing status one can simply study all ethnic households who move within the city versus those who move to suburban locations, and ask whether the latter are more likely to be homeowners. This proxy analysis of immigrants' moves in fact emphasizes the difference between recent immigrants, who are less likely to be able to move to ownership, and earlier immigrants who have acquired the capital to make the transition.

The analysis focuses on Asians in the Bay Area and on Hispanics (in total), Mexicans, and Asians in Southern California. The data are drawn from the 1990 Public Use Microdata Files and are limited to households that either changed location within the city or moved from the city to the surrounding counties. I plot the rate of homeownership for all movers *within* the city against the rate of ownership for all movers from the city to the county. The rates are computed for Asians for Northern California and for Asian, Hispanics, and Mexican ethnicities for Southern California (Figures 5.7 and 5.8). The contrasts between intracity movers and those who move to the more suburban counties are striking.

When we examine the figures we must also note that moves within Los Angeles City can certainly be to "suburbanlike" settings in the San Fernando Valley, but the tendency toward increased ownership in suburban moves would be greater if the San Fernando Valley was included as a "suburban" destination. Significant differences in homeownership rates for inner-city moves and for movement out confirm the importance of moving out and moving up, statistical confirmation of the commentary in Abelmann and Lie (1995).

Among Asian immigrants moving into dwellings within the City of San Francisco, the rate of homeownership is about 40%. Among their counterparts destined for Alameda, Contra Costa, and other suburban San Francisco Bay communities, the corresponding rate is nearly twice that level. Clearly the higher ownership rates of those who move out reflect greater length of residency and larger incomes; nevertheless, they are strik-

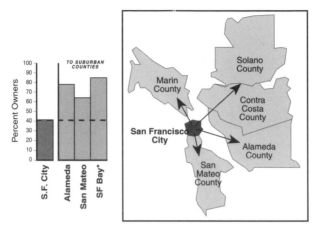

FIGURE 5.7.   Trends to homeownership for suburban moves in Northern California. *Source:* U.S. Bureau of the Census, 1992.

ing evidence that the dream of ownership and suburban location is alive and well among immigrants.

The differences are no less striking in Southern California. Ownership rates for those who relocate outside the City of Los Angeles are uniformly higher—often as much as three times higher. Obviously, the rates for movers within the city are much lower because of the large number of new immigrants who possess few assets, have low incomes, and are unable to enter the homeowner market. Just as obviously, however, those who can accumulate the assets, increase their income, and move are becoming homeowners. The rates of ownership for Hispanics generally and for Mexicans specifically are lower than those for Asians overall and for Asians in the Bay Area, but in earlier chapters I established that there are differences between the regions and that the ownership rates are simply extensions of these economic differences.

The number of movers is small; even so, the path of the white suburban migration is viable, at least for limited numbers of the new ethnic groups in California. As always the question is about the sustainability of what is still a fledgling process.

IMMIGRANTS AND RESIDENTIAL CROWDING

On the less bright side of the immigrant housing picture, many immigrants face severe constraints on housing affordability, which force doubling up,

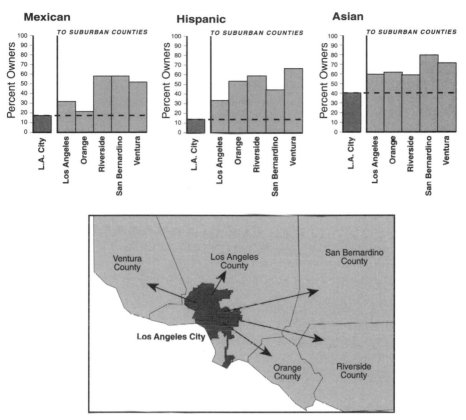

FIGURE 5.8. Association between suburbanward movement and homeownership by immigrant group, 1985–1990, in Southern California. *Source:* U.S. Bureau of the Census, 1992.

illegal conversions, and deferred maintenance and upkeep on existing units. We have increasing evidence that just such housing processes are occurring in inner-city immigrant neighborhoods.[2]

In Los Angeles the number of high-poverty tracts increased by 79% between 1970 and 1990. The number of people in those tracts increased by 170,829. These high-poverty neighborhoods often provide the housing for new immigrants. In San Francisco, however, the high-poverty tracts did not increase and the number of persons in those tracts increased by only 10,000. The issue of poverty and crowding seems to be one that is more real in Southern California than in the Bay area, although on the local scale (as I will demonstrate) both regions have concentrations of crowded immigrants.

The Los Angeles metropolitan area includes a band of crowded, largely immigrant and poor inner-city neighborhoods east of downtown, but other concentrations exist throughout the region. The housing in high-poverty neighborhoods in California, however, does not fit Jargowsky's (1996) description of abandoned and shuttered housing with significant evidence of neighborhood disinvestment. Residential buildings are not vacant or abandoned. The housing may be of poor structural quality, but it does not lack kitchens, plumbing, or toilets. Rather, the problem is one of the stress created by very crowded conditions and the resulting burden placed on neighborhood services.

Many immigrant neighborhoods are severely crowded. In the Salvadoran neighborhoods near downtown Los Angeles, where about three-quarters of the population are foreign-born, some population densities reach almost 70,000 persons per square mile. We would expect to find such densities more commonly in the inner cities of India and Africa than in North America. Half the population lives in very crowded housing[3]; one-fifth live in units containing 6 or more persons per unit. Crowding and densities are almost as high in the Cambodian neighborhoods in Long Beach, but not quite as high in East Los Angeles (Figure 5.9).

Even though the aggregate data suggest that poverty and crowding are less severe in inner-city neighborhoods in San Francisco, at least some inner-city Asian neighborhoods in San Francisco are as crowded as any neighborhood in Los Angeles. The Asbury neighborhood in San Francisco is over 90% Asian and more than 70% foreign-born, with very high population densities but somewhat lower crowded conditions (Figure 5.9). The crowding is not as great in immigrant concentrations in San Jose and Redwood City, though nearly one-third of the neighborhood population in San Jose and one-fifth of the population in Redwood City live in very crowded conditions.

In the long run, the adequacy of the housing may obviate the negative effects of crowding and high density. At the same time, the existence of such conditions in immigrant enclaves confirms how difficult it is for at least some of the immigrants to escape the concentrations of poverty and crowding, and demonstrates the "spillover" effects of poor community services and neighborhood decay. While some observers celebrate the vibrant ethnic diversity of inner-city neighborhoods, others see a darker side to these very dense concentrations of new and often illegal immigrants. Not the least of the problems is how the new immigrants in such settings can enter the educational system.

EDUCATING THE NEW IMMIGRANTS

To the new immigrants, ensuring that their children have the best chances is as important as owning a piece of the new society. Surveys consistently

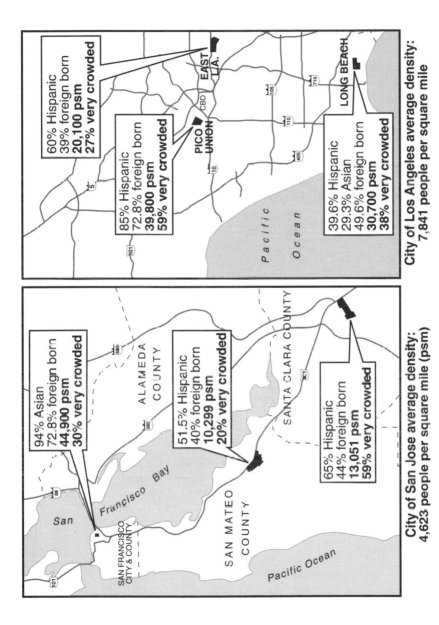

City of San Jose average density:
4,623 people per square mile (psm)

94% Asian
72.8% foreign born
44,900 psm
30% very crowded

51.5% Hispanic
40% foreign born
10,299 psm
20% very crowded

65% Hispanic
44% foreign born
13,051 psm
59% very crowded

City of Los Angeles average density:
7,841 people per square mile

60% Hispanic
39% foreign born
20,100 psm
27% very crowded

85% Hispanic
72.8% foreign born
39,800 psm
59% very crowded

39.6% Hispanic
29.3% Asian
49.6% foreign born
30,700 psm
38% very crowded

FIGURE 5.9. Housing crowding in selected neighborhoods in the Bay Area and Southern California. *Source:* U.S. Bureau of the Census, 1993.

113

emphasize the importance of education for recent arrivals. Although the data on college aspirations varies by ethnic background, it is revealing that immigrant parents for the United States as a whole have higher aspirations for their children than do native-born parents as a whole (Figure 5.10).[4] Immigrant parents also differ significantly in their aspirations. Hispanic parents have the lowest aspiration rates, though again they are higher for immigrant parents than for native-born parents.

At the same time, the statistics tell a more complex story about educational success. Overall, the educational gains are notable, but not all groups are doing equally well, and some stresses are involved in delivering an adequate education to the new immigrants.

A recent national study reports that one-third of young Hispanic adults were high school dropouts in 1995 and that the drop-out rate has not changed for two decades (U.S. Department of Education, 1997). This finding is a critical thread in the immigrant story. Very high drop-out rates reduce immigrants' "human capital," especially that of Hispanic immi-

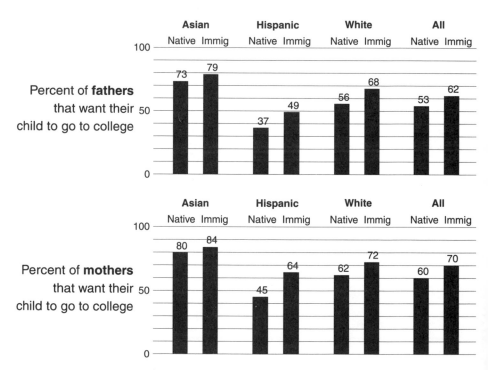

FIGURE 5.10. Parental expectations for high school seniors (U.S. data): Percent who want their child to go to college. *Source:* Vernez and Abrahamse, 1996.

grants, who constitute the major immigrant group in California and are the group most likely to drop out. The evidence suggests that without significant increases in the educational levels of this group, at least a large segment of the new immigrants will not be able to realize the dream of upward social and economic mobility.

It quickly becomes apparent that future trajectories of immigrants' success are closely tied not only to the immigrants' own paths, but also to the paths of their native-born children. Thus the questions that I examine here concern the issues and impacts that will arise from the social and economic future of the citizen children of recent immigrants' as well as of immigrants themselves. The study of immigrant children and their educational success is important because (as others have also observed) the economic penalties of not adequately educating new migrants, and of failing to incorporate them into mainstream U.S. society, will be far greater than the costs of educating this population properly (Bean et al., 1994, p. 95).

Educational Levels and Trajectories

Immigrants vary widely in educational achievement. Some hardly differ from the native-born populations; others are much less likely to receive substantial years of education. The greatest differences are between those immigrants from Central America and Mexico and immigrants from Asia, but even the Asian immigrant populations show large educational differences (Table 5.4).

Overall, the proportion of immigrants without high school graduation diplomas includes a very large number of Latino migrants from Mexico and Central America, where the opportunity for education beyond the primary levels is quite limited. As Table 5.4 shows, almost 70% of immigrants from these regions lack a high school education. In contrast, the Middle Eastern immigrants, as a group, are almost the exact reverse of the

TABLE 5.1. Educational attainment by immigrant status in California in 1990

Education	All	Mexico/ Central Am.	Vietnam/ Cambodia/ Laos	Other Asia	Middle East
Not high school grad.	42.7	69.4	46.8	20.6	8.4
High school grad.	19.3	18.1	23.7	18.6	19.9
Some college	16.0	8.6	19.6	22.2	11.2
College degree plus	22.0	3.9	9.9	38.7	60.6

Source: U.S. Bureau of the Census, 1992.

Latino migrants. Among Asians the level of education is mixed: some Asian immigrants come from countries with highly developed educational infrastructures, and others come from countries with weak educational systems. The findings for Mexican and other Latino immigrants are not surprising. They largely reflect the fact that many immigrants come as young adults at the beginning of their working lives and are unlikely to pursue more education.

In general the level of education of immigrants is holding steady or increasing. The percentage of the immigrant population with 8 or fewer years of education was much lower in 1990 than it was in 1965 (Simon and Akbari, 1995). In addition, the proportion of immigrants with 16 or more years of education increased over the same period. It appears, too, that immigrant children are as likely as native-born children to enroll in U.S. primary and middle schools. Even more important, immigrants who do go on to high school in the United States as a whole are as least as likely as natives to graduate and are more likely to aspire to a college education (Vernez and Abrahamse, 1996).

All immigrants as a group had an average of just less than 11 years of education in 1970, but slightly more than that amount in 1996 (Table 5.5). The average years of education for recent immigrants (those who entered in the 5 years before the respective census year) also increased from 10.4 years in 1970 to 11.3 in 1996, although (as the table shows) the average fluctuates around a mean of 11 years. We find varying trends by broad ethnic group: even though the Mexican and Central American recent immigrants are arriving with more years of education, they start at very low levels of educational achievement. In addition, the average years of education mask more detailed ethnic differences. For example, while 70% of Filipino immigrants had at least some college education, only 3% of Salvadoran immigrants had reached this level (DaVanzo, Hawes-Dawsan, Valdez, and Vernez, 1994).

TABLE 5.5. Average years of education for all immigrants and for recent immigrants (Arrived in the last 5 years)

	1965–1970	1975–1980	1985–1990	1990–1995
All immigrants	10.8	11.3	10.8	11.2
Recent immigrants	10.4	11.2	10.6	11.3
Asian recent immigrants	13.5	13.2	12.5	12.1
Middle Eastern recent immigrants	—	11.9	12.5	15.9
Mexican recent immigrants	6.9	7.4	7.8	8.4

Source: U.S. Bureau of the Census, 1972, 1983, 1992.

In any discussion of the success of the new immigrants the big question is how well the Hispanic population, immigrant and nonimmigrant alike, is doing, both in staying in school and in achieving more advanced levels of education. It is a matter of great concern that many native-born Hispanic Americans aged 18–21 have low "in-school" rates and the lowest college completion rates (Figure 5.11). One-half of all Hispanic high school graduates go on to college (one-third of the students in California high schools are of Hispanic origin), but four out of five Asian high school graduates do so. Again, we see how much the findings are influenced by the particular ethnic origin.

Unfortunately, the levels of college education for Hispanics are low and persist across age groups. In contrast, non-Hispanics continue to increase their education: 27.6 of the 25 to 29-year-old group have a bachelor's degree or higher, and 36% of the 40–44-year-olds, an increase of almost 10%. The same comparison for college-educated Hispanics shows only a modest increase with increasing age, from 6.8% to 9.0% (Figure 5.12). Hispanics not only have very low college completion rates, but also very high noncompletion rates for high school. About 50% of Hispanics do not finish high school and this rate increases with age (Figure 5.12). The proportion of Hispanics who do not complete high school ("high school dropouts" in educational terminology) is much higher than that of either black or white non-Hispanic students (Figure 5.13). The rate has fluctuated considerably, possibly as a result of differing immigration surges and IRCA legalizations. In spite of the fluctuation, the rate generally declined until 1993.

Hispanic students' high drop-out rates are explained in part by whether the Hispanic students were ever in the system. It seems that young

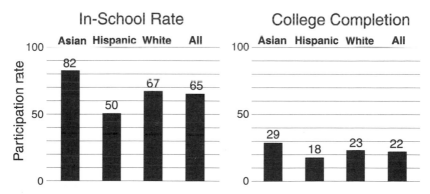

FIGURE 5.11. Proportions currently in school for 18–24-year-olds and college completion rates. *Source:* Vernez and Abrahamse, 1996.

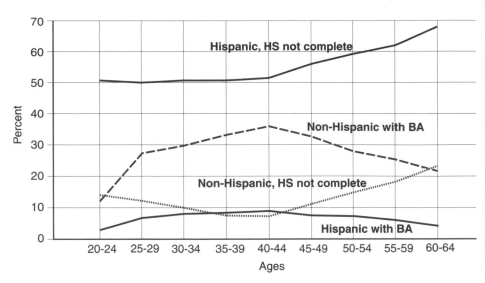

FIGURE 5.12. Education levels for Hispanics and non-Hispanics, 1990. *Source:* U.S. Bureau of the Census, 1993.

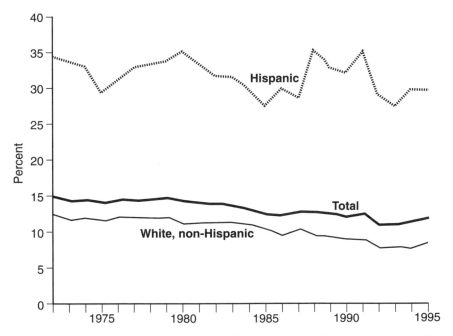

FIGURE 5.13. Drop-out rates for grades 10–12, ages 15–24, by race/ethnicity. October 1972 through October 1995. *Source:* U.S. Department of Education, 1997.

Mexican migrants who have already been out of school 2 years by age 15 do not enroll in U.S. schools, either by choice, because of their inability to catch up with others of their age, or from economic necessity (Vernez and Abrahamse, 1996). Among all foreign-born Hispanic youths who ever entered the U.S. high school system, the completion rate is about 54%; in other words, 46% dropped out. Of the young adults who were counted in the Current Population Survey of 1995 and who were born outside the United States, almost half had not completed high school.

Of all Hispanic immigrants to the United States (and approximately 60% of these live in California), about 43% had never enrolled in a U.S. high school (Figure 5.14). That is, many immigrants arrive with limited education and fail to acquire any more. The drop-out rate for Hispanic students who did enroll in a U.S. high school was a little under 20%; for foreign-born Hispanic students, it was approximately 24% (U.S. Department of Education, 1997). Although this rate is high, it is much lower than the 30% rate for Hispanics in Figure 5.14.

Summing up, about one-third of the dropouts consist of young Hispanics who come to the United States without a high school education and never enter U.S. schools. This low level of education directly influences their life chances. Those without a high school education who never enrolled in U.S. schools are particularly disadvantaged: only half of all for-

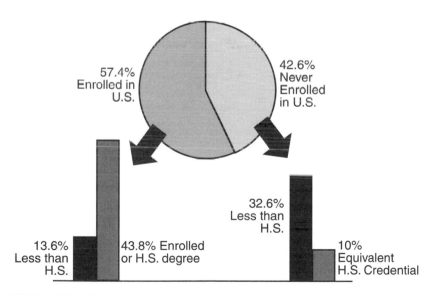

FIGURE 5.14.   Hispanic immigrants, ages 16–24, by high school education status. *Source:* U.S. Department of Education.

eign-born Hispanics aged 16 to 24 who did not enroll in a U.S. school completed a 7th- or 8th-grade education, but for native-born *and* foreign-born Hispanics who enrolled but did not finish high school, more than 90% did gain a 7th- or 8th-grade education.

Overall, we know that immigrants, especially Hispanic immigrants, have lower-than-average levels of educational achievement. However, the key question is What is the future trajectory of these immigrants and their children? We find some evidence of a positive trajectory, though again there are mixed messages about the rate at which education is acquired. Some studies paint a picture of a successful transition to native-born or even higher levels of education; others are less sanguine. Throughout all the studies, the difficulties of the Latino population are a constant theme. Kao and Tienda (1995) and Rumbaut (1997) portray a picture of the successes of immigrants' adaptation to learning. Both studies emphasize the success of the children of immigrants rather than immigrant children themselves. Generation status was important in shaping educational outcomes. First and second generation students in general performed as well or better than their native-born counterparts, and parents optimism about their children's educational prospects was an important part of the children's achievement (Kao and Tienda, 1995). The results, however, were less compelling for the children of Hispanic immigrants.

Rumbaut's recent study of immigrant children in San Diego finds positive educational outcomes for the children of immigrants. The San Diego study, which is most relevant to our discussions of California communities, followed 2,400 students from 8th and 9th grade to almost the end of high school. These students had at least one foreign-born parent, and in general they did as well as native-born children, or better. Forty-four percent of the 9th grade students with immigrant parents had a grade point average of 3.0 or higher, in comparison with 29% of native-born children. The gap still existed in the 12th grade, though the difference then was only 4%.

Despite the considerable variety within the native-born population, the point of the study is a good one: immigrant children are doing better than native-born children in some situations. Not only were students doing better, they were less likely to drop out. The differences between native-born Asian students and immigrant Asian students are smaller than for Hispanic students: only about 5.8% of the native-born Asian students and 4.5% of the immigrant Asian students were likely to drop out. The drop-out rate for Hispanic immigrant children was 8.5%, compared with 26.5% for native-born Hispanics (Rumbaut, 1997).

These findings are reassuring, but a detailed examination of the data on performance is not always so encouraging. Standardized reading and mathematics scores are lower for immigrants than for native-born students; for Mexican immigrants, they are very low (Figure 5.15). The scores are higher for Asian immigrant children generally, but low for Cambodian and high for "other immigrants."

The findings are both positive and cautionary. The cautionary findings again emphasize the challenge to California society posed by the influx of Hispanic students. As always, the question is how to educate these students and how to integrate them into U.S. society. The speed and the extent to which these immigrants will acquire education is dependent on the conditions for learning, especially the extent of poverty, which is one of the most important constraints on learning.

UNDERSTANDING THE LINK BETWEEN POVERTY AND LEARNING

The research in the last half-decade has identified family income as an important influence on children's achievement levels. Several contemporary studies have shown that low income in early childhood consistently leads to low levels of achievement in later education. Parental income has a positive, significant effect on children's later success in the labor market. Haveman and Wolfe (1995) conclude, in their review of several studies, that growing up in a poor family appears to have a particularly negative effect on later success in the workforce and on earnings. These studies also show that family income in early and middle childhood is far more important for shaping ability and achievement than income during the adolescent years. In addition, the effect of income on completed schooling—that is, the years of schooling a child achieves—is the most negative for children in low-income families (Duncan, Brooks-Gunn, and Klebanov, 1994; Haveman and Wolfe, 1995).

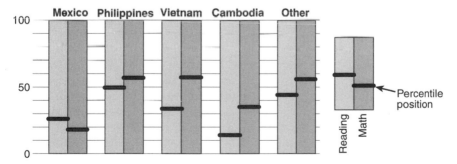

FIGURE 5.15. National percentile rank for reading and math for foreign-born San Diego schoolchildren. *Source:* Rumbaut, 1997.

Survey data provide additional information on early childhood experiences as they relate to school readiness. A study of 4,423 children conducted in 1993 showed that mother's education has important effects on early signs of literacy and numeracy (Zill and Collins, 1995). The emphasis on low maternal education and minority language status, rather than on income, as strong predictors for the level of children's development and accomplishments is not inconsistent with the findings that low income affects later achievement levels, but it enriches our understanding of how learning is accomplished, and how the human capital of education is produced. Clearly, this is a complex process in which the social and economic status of the early childhood home are critical factors.

Specific studies of the link between poverty and achievement have also shown place and locality effects. Attending a school with a high concentration of poor students is strongly associated with low achievement in reading and mathematics (U.S. Department of Education, 1993). Overall, there is a strong consensus that low incomes carry higher risks of reduced access to education, information, and training (Weisbrod, 1965). It is also reasonable to argue that parents in poverty will find it much more difficult to give their children better healthcare and education, and the associated skills needed to improve their chances in society.

If poverty influences the trajectories of success for the native-born population, it is likely that this pattern is intensified for immigrants and for the children of immigrants, who are even less skilled and have fewer years of basic education. The problem is further compounded by a context in which job opportunities are decreasing and in which occupational success requires increasing skill levels. Not surprisingly, and especially important for this study, is the general finding that poverty is greater among black and Hispanic households (Sawhill, 1988) and thus, by extension, among immigrant households.

Lower achievement levels intersect with changing occupational opportunities. Although low-paying service-sector jobs are available, these jobs do not provide upward trajectories for the new immigrants. Some observers argue that the immigrants will follow the path of earlier immigrants, adapt successfully to their new environment, and eventually participate fully in the host economy. These observers see a convergence of immigrant and native-born characteristics. Yet if the old paths are no longer available and if the new immigrants do not acquire more than a minimum education, the rungs on the ladder of opportunity will be more difficult to climb, and the pattern of the past cannot be repeated. Myers (1996) draws an important distinction in the path of assimilation between those who tended to learn English rapidly and become citizens, and others with much lower levels of language skills. Thus poverty status and language skill are important, if not critical, elements of the learning context.

I have already shown that poverty in very young children is high for Hispanics and has increased substantially in the last decade. How does this poverty vary by the parent's citizenship status and for the school-age population? How does it affect the educational outcomes of the two most vulnerable groups, Mexican and Central Americans, and Southeast Asians?

As elsewhere in the book, I use the Public Use Microdata Sample (PUMS) for California for 1990 to differentiate the native-born from the foreign-born population and to distinguish the citizenship status of the foreign-born population. I am interested specifically in the children of the foreign-born population and their poverty status. If they grow up in poverty, either in absolute terms or in relation to the native-born population, there is likely to be a significant affect on their learning curve. I focus here on the current child population aged 0–17. That is, children age 0–17 have parents who are approximately 20–40 years old, and who entered the United States either as children or as young adults.

Some 7.7 million children aged 0–17 lived with parents or with other householders in California in 1990, and an additional 0.3 million lived in group quarters. The analysis focuses on the children not in group quarters, of whom about 5.2 million were the children of the native-born population (Table 5.6). The remaining 2.5 million children (almost 33%) had at least one foreign-born parent. Most of these were citizen children of foreign-born parents.

There are very large differences between the poverty status of the children of the native-born population (even without racial and ethnic differentiation) and the poverty status of the children of foreign-born parents. While 14.4% of the California native-born child population are in poverty, close to twice as many, 24.8%, of the children of the foreign-born occupy this status. The breakdown by citizen status and origin causes even greater concern about the potential effects for a poverty population. On average, one-quarter of the children of the foreign-born population are in poverty, but the proportion for Mexican and Central American children of the foreign-born population varies, by the children's citizen status, from almost one-third to nearly 40%. For the children of refugees (largely from Vietnam, Cambodia, and Laos) the proportion is close to half, and for noncitizen children it reaches 57% (Table 5.6).

If we accept the thesis that income and (by extension) poverty are critical dimensions of children's later life-course success, the data for California suggest a serious problem. Even the most successful educational program will find it difficult to educate noncitizen children, many of whom lack language proficiency as well as suffer the problems of poverty. Poverty is a major dimension of the problem of creating an educated population; language acquisition is the other.

TABLE 5.6. Children by parental citizenship status and poverty in California*

	Citizen children	%	Mexican, Central Am.	%	Vietnam, Cambodia	%	Other	%
All children								
Parent native-born	5,185,286							
Parent foreign-born	2,542,074		1,431,004		210,640		900,420	
Citizen	1,738,676		1,013,051		107,097		618,528	
Naturalized	108,002		47,772		12,679		47,551	
Not citizen	64,728		354,646		89,674		198,408	
(N/A)	52,668		15,535		1,190		35,943	
Total	7,727,360							
Children in poverty								
Parent native-born	745,230	14.4						
Parent foreign-born	630,716	24.8	427,717	29.9	98,639	46.8	104,360	11.6
Citizen	366,097	21.1	270,377	26.7	43,979	41.1	51,741	8.4
Naturalized	26,151	24.2	16,818	35.2	3,254	25.7	6,079	12.8
Not citizen	231,658	36.0	137,116	38.7	51,163	57.1	43,379	21.9
(N/A)	6,810	12.9	3,406	21.9	243	20.4	3,161	8.8
Total								

*Poverty not defined for 153,376 native-born and 24,203 foreign-born.
Source: U.S. Bureau of the Census, 1992.

Language

English language acquisition and use are critical in facilitating learning and in acquiring training, which will lead in turn to successful participation in the labor market. Acquiring English is even more important in an information-based postindustrial society. Immigrants traditionally arrived in the United States with limited English proficiency and slowly gained language skills with length of residence. The long period of low immigration created a sense of language homogeneity; those who did not speak English were regarded as people in transition. Now, after several decades of increased immigration, the question of language use and acquisition again occupies center stage. In California, 31.5% of the population do not speak English as their first language. Only 8%, however, report that they do not speak it at all or do not speak it well (U.S. Census of Population, 1990). The issue is that those who do not speak English or who do not speak it well are correlated with lower income, less-well-educated immigrant populations, and recent immigrants who are mostly refugee or Latino groups.

Language use and acquisition also affect the nature and rate of assimilation, which I examine in Chapter 6. The concern in this chapter is the role of language in learning. I have already demonstrated that the number of students with "limited English proficiency" has increased dramatically in recent years, but some facts are worth restating. The number of LEP students increased by 42% between 1985 and 1995; LEP students now represent about one-quarter of all students in the California school system. Spanish was the primary language of 78% of all LEP students. In other words, the language acquisition problem is most acute in the low-income, less-well-educated immigrant groups. It is precisely these groups which find it difficult to make the transition to the high-technology labor market and, when they do acquire jobs, to move up though the labor market.

The research shows that language affects labor-market participation and earnings. The native-born gain their skills in the "mother tongue" in the early years of caregiving, and continue to enlarge their language skills through the years of formal schooling. For the foreign-born, the acquisition of language skills is more difficult, more time-consuming, and more costly. Chiswick and Miller (1996) identified three factors that influence the acquisition of language capital: exposure to the language, efficiency in language acquisition, and economic incentives for language acquisition. Exposure is influenced by duration of residence. It is also affected by marriage to a spouse who speaks a different language, which decreases exposure to the native language and increases exposure to the new language, versus marriage to a spouse who speaks the same language, which decreases exposure to the new language (Evans, 1986). Efficiency refers to the extent to which a given amount of exposure is translated into skills in the new language: the young are very efficient in making this transition. Finally, those who are only at the beginning of their labor-market participation have a greater incentive to acquire language skills. They will be able to use them for a longer time and will gain greater rewards (Krashen, 1982).

Does acquiring more language skills pay off? Chiswick and Miller (1995) report that for immigrants to the United States as a whole, fluency in English will raise earnings by almost 17%. The greatest gains in fluency are made by those with longer residence, by groups in which a smaller proportion speak the original language, and in marriages in which the spouse has a different mother tongue. These findings are particularly important because they hint at the problems facing the large Latino population in California, who are close to the source of their native language and who generally marry other Spanish speakers. In this context, language acquisition is slower and the children will acquire English less quickly. This is another facet of the limitations on acquiring human capital.

Is Bilingual Education a Solution?

There is no question that a large proportion of California's children have LEP. The county maps in Chapter 3 show that this population is concentrated in Southern California and in the rural agricultural counties (see Figure 3.17). In response to the problem of increasing numbers of students with inadequate English-language skills, the schools have emphasized teaching in those students' native language in order to bring them up to speed in English. Has this focus on bilingual education worked? Is it a solution?

There are more than a million students enrolled in bilingual education programs in 7,093 schools in California; they constitute about 20% of all students in the California school system. A majority of these are in elementary schools (Table 5.7). The system includes 12,100 teachers who provide instruction in the primary language, 15,765 teachers who provide English-language development, and an additional 29,000 bilingual paraprofessionals. Almost 30,000 teachers are in training for bilingual teaching (Education Demographics Unit, California Department of Education, 1995). This bilingual education costs approximately $300 million a year. Overall the growth in bilingual education matches the growth in LEP students.

The Bilingual Education Act of 1968 was designed to address the needs of students who were a "linguistic minority." The act was motivated by Hispanic students' lower achievement in comparison with white children and by the common practice of forcing Spanish-speaking students to learn English by placing them in regular classrooms where all the instruction was delivered in English. Amendments to the act in 1974 and 1978 provided for the use of the children's native language to help them become competent in English. Further amendments and revisions in 1984 and 1988 encouraged more flexibility for state and local school districts, and specifically allowed different strategies for educating LEP students.

In bilingual education, non-English-speaking students are taught to

TABLE 5.7. Distribution of students receiving bilingual education in California in 1996

Grade	No. students	%	Ethnicity	No. students	%
Elementary	835,734	66.2	Spanish	990,801	78.4
Middle	196,641	15.6	Vietnamese	48,907	3.9
High school	223,568	17.7	Hmong	30,345	2.4
Other	7,039	0.6	Cantonese	23,954	1.9
			Other	168,975	13.4
Total	1,262,982	100.0	Total	1,262,982	100.0

Source: Language Report of the Demographic Unit.

read and write in their native language, learn skills in their native tongue, and move slowly from their native language to all-English instruction. This program is quite different from the educational practice in Europe, where schools provide supplementary and often intensive teaching in the language of the host country for 1 or 2 years and then move children as quickly as possible to regular classroom instruction in the host-country language. Bilingual education in the United States and in California is the subject of a contentious debate.[5] On the one hand, critics of transitional bilingual education argue that children come out of the program not knowing English. On the other hand, the bilingual education supporters argue that teaching only in English limits the LEP child's intellectual development and damages his/her self-confidence and self-esteem (Rossell and Baker, 1996).

Many studies of bilingual education express support; many others oppose the practice. The consensus, however, appears to be that bilingual education neither harms nor helps (Rossell and Baker, 1996). The question, of course, is whether bilingual education meets its goal of producing increased achievement and the transition to high achievement in English and in subjects tested in English. Rossell and Baker reviewed 72 studies that statistically evaluated the achievement differences between students who were taught in bilingual programs and those who were not (matching the students properly). They found no evidence supporting the superiority of transitional bilingual education. Other studies (Baker and de Kanter, 1983; Holland, 1986; and Rotberg, 1982) previously had reached a similar conclusion. Of even greater concern, the studies reviewed by Rossell and Baker provide convincing data supporting some form of immersion program to teach English and other subjects.

If there is little hard evidence in support of bilingual education, why is it advocated? Rossell and Baker suggest that the thoughtless treatment of linguistic minorities (punishing students for using their native tongue), low achievement levels, and high drop-out rates have convinced a diverse group of social scientists, educators, and civil rights attorneys that it is more equitable and less racist to be sensitive to native languages. In such a climate it is difficult to criticize such programs and easy to categorize attempts to evaluate these programs as racially motivated. But what is the evidence for the success, simply in the aggregate, in California? Have differences in test scores narrowed over the decade in which large numbers of students have been tutored in bilingual programs? Moreover, some Hispanic parents are increasingly vociferous in their rejection of bilingual education: they feel that their children are not learning English and are being disadvantaged in their future employment opportunities (*Education Week* on the Web, 28 February 1996).

The data for California are equivocal at best. The latest figures show a

2.1% increase in seniors scoring at or above national averages on the SAT and ACT college admissions tests. But it is only when we examine the absolute levels that we can view these results with real concern. The percentage of seniors in all of California scoring at national norms was 19.5% in 1995. This is hardly an endorsement of either the educational system, the bilingual education system, or the process of education in the state. In addition, college attendance decreased in the period from 1989 to 1995. The proportion of high school graduates attending California public 2- and 4-year colleges has declined by a little more than one percentage point (California Department of Education, 1997). These average data mask the figures for the inner city, where the schools are full of immigrants, and where the percentage of students scoring at or above the national average is barely on the scale.

It is widely believed that bilingual education has improved educational outcomes. Yet the evidence, though controversial, seems to indicate the opposite. At the same time, the evidence from a study of Head Start programs for Latinos provides substantial evidence that early intervention brings significant benefits (Currie and Thomas, 1997). On average, participation in the Head Start program closes about one-quarter of the gap between Latino and non-Hispanic white children and two-thirds of the gap in the probability that young children will repeat a grade. In addition, the gains are greatest for children of Mexican origin, especially children with native-born mothers. Clearly, we are revisiting issues of human capital; the children of mothers with more human capital do significantly better than those with less human capital. These findings are important because they bolster the arguments about early childhood deprivation that I examined earlier in this Chapter . Children need help early and the data from the Currie and Thomas (1997) study emphasize that finding.

WHY IMMIGRANT POVERTY MATTERS
FOR EDUCATIONAL OUTCOMES

To make a plausible argument that family and child poverty is already affecting achievement levels, I must demonstrate the existence of poverty in the immigrant population and show the real effects of poverty on immigrant childrens' educational success. It is relatively straightforward to demonstrate the first point, but the issue of effects is more difficult. In this case study, it relies on an examination of achievement levels in schools and school districts in Los Angeles County and the San Francisco Bay area. Although we cannot link individuals across the poverty/achievement scale, poverty is clearly associated with low achievement in the heavily immigrant school districts. One can argue that the low achievement levels are a temporary outcome of immigrants' poverty and low incomes, but the evidence raises unavoidable questions about the short-term versus the long-term implications of the links between poverty and success. Yet, because these achievement scores are for 11th-grade students, that is, for high school students who will graduate soon and enter the labor force, the likelihood of success is limited at best.

Los Angeles County contains 88 independent cities and a nearly equal number of school districts, of which Los Angeles Unified is the largest. The Los Angeles Unified School District, the largest in the county and the second largest in the nation, enrolled 632,804 students in 1993–1994. Other school districts in the county have enrollments ranging from 3,000 to more than 100,000 students. The Los Angeles district is not only the largest but also the most diverse, although the continuing increase in Hispanic students is reducing that diversity. Currently, the Los Angeles Unifed School District is majority Hispanic with 66% Hispanic students, 14% black, 12% non-Hispanic white, and 8% Asian.

The schools and school systems in California and Los Angeles County have been affected by the very large-scale recent immigration. Los Angeles County received a very large proportion of all the new immigrants to California, many of whom are LEP. In the Los Angeles Unified School District, 57% of the students are classified as LEP. It has been suggested that the large-scale recent immigration, combined with the subsequent increase in citizen children who are in poverty and who are also LEP students, will provide a critical cluster of long-term problems for the large metropolitan areas of California. Thus an analysis of Scholastic Aptitude Test (SAT) scores, specifically the percentage of students in a school or school district who are above the national average SAT score; school completion rates; and the proportion of LEP students will show the relationship between outcomes and contexts. It is possible to identify poverty populations by the proxy figures for children receiving free and reduced-price lunch support in each school and each school district. The approach used in this chapter reveals a context in which immigration, poverty, and low levels of achievement overlap.

Poverty is higher in Los Angeles County than in California as a whole (Table 5.8). While 14.4% of the state's native-born population live in poverty households, the percentage is slightly higher in Los Angeles County. Similarly, 26.3% of the foreign-born population in Los Angeles County live in poverty, a proportion higher than that for the state as a whole. The children who are naturalized or who are not citizens are significantly more likely than citizen children to be living in poverty. In Los Angeles as a whole, the rates of the children in poverty increase for noncitizen children and for Mexican/Central American and Vietnam/Cambodian citizen children. In some cases, nearly half of the children with foreign-born parents

TABLE 5.8. Children by parental citizenship status and poverty in Los Angeles County*

	Citizen children	%	Mexican, Central Am.	%	Vietnam, Cambodia	%	Other	%
All children								
Parent native-born	1,197,697							
Parent foreign-born	1,128,076		745,020		5,122		337,934	
Citizen	766,203		530,239		21,751		214,213	
Naturalized	50,045		26,702		3,434		19,909	
Not citizen	297,553		182,225		19,742		95,586	
(N/A)	14,275		5,854		195		8,226	
Total	2,325,773							
Children in poverty								
Parent native-born	205,398	17.1						
Parent foreign-born	296,385	26.3	228,951	30.7	18,241	40.4	49,193	14.6
Citizen	174,749	22.8	146,949	27.7	6,889	31.7	20,911	9.8
Naturalized	13,480	26.9	9,420	35.3	1,109	32.3	2,951	14.8
Not citizen	105,602	35.5	71,446	39.2	10,150	51.4	24,006	25.1
(N/A)	2,554	—	1,136	—	93	—	1,325	—
Total	501,783							

*Antipoverty not defined for 44,321 native-born and 12,374 foreign-born.
Source: U.S. Bureau of the Census, 1992.

are in poverty. Poverty rates differ, depending on the origin of the foreign-born population. In addition, more recent arrivals, the children of the foreign-born who are not citizens, are doing less well; more of these children are in poverty households. One cannot escape the fact that large numbers of citizen children in Los Angeles County are growing up in poverty.

My central argument is that poor scores on achievement tests are the result of very high levels of poverty and reflect the existence of large numbers of LEP students. The mathematics and reading scores overall, and particularly for LEP students (if we recall the argument that LEP students are almost certainly immigrants or the children of recent immigrants), clearly indicate the joint problem of recent mass migration and poverty. An examination of Figures 5.16 and 5.17 suggests the magnitude of the problem facing the immigrant ports of the United States. No school in the sample that is in or near an immigrant concentration reaches the national norm in reading scores, and only one surpasses the national norms in mathematics scores. LEP students' scores are half of those of all students in reading, but

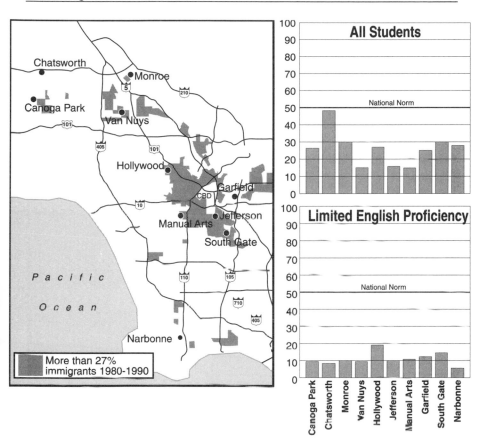

FIGURE 5.16. Reading scores for selected Los Angeles high schools, 11th grade, 1995–1996. *Source:* Los Angeles Unified School District.

only one-third lower in mathematics. While 11th-grade LEP students are performing at about one-fifth of the national level for reading, they are performing on average at about half the national norm for mathematics scores in this sample of schools. It is hard to argue that such a pervasive pattern is simply a school district effect. Los Angeles is a microcosm, in which we are witnessing a serious problem in the ability of inner-city school systems to receive and educate the flows of new immigrants.

It is relatively simple to document the extensiveness of poverty and the low achievement scores in Los Angeles County, but more complex to establish a link between poverty and school performance. Even though we can use only aggregate poverty data and average scores, recall (as I argued above) that children suffer the effects of both growing up in poverty and of

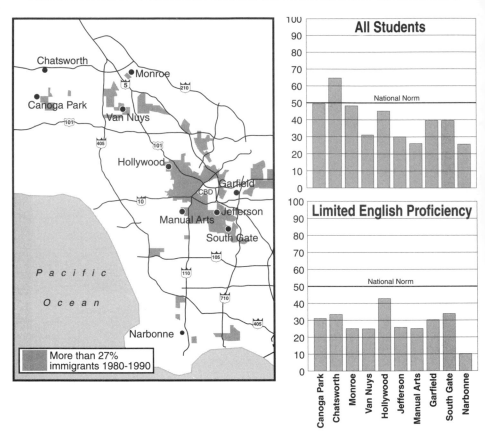

FIGURE 5.17. Math scores for selected Los Angeles high schools, 11th grade, 1995–1996. *Source:* Los Angeles Unified School District.

attending a school where a large proportion of the children are poor. As a first step in demonstrating the links between poverty and achievement, this analysis uses overall test outcomes in selected schools in the Los Angeles Unified School District and in a random comparison set of school districts in Los Angeles County, as well as corresponding data on poverty, as measured by how many students receive free and reduced-price lunches (Table 5.9).

Los Angeles County is home to almost one-third of a million children in poverty and with foreign-born parents. More than a quarter-million children in poverty are Mexican and Central American in origin. Most of these parents and their children are in the Los Angeles Unified School District and clearly account for a large proportion of the more than 400,000

TABLE 5.9. Education levels and outcomes in Los Angeles County, 1995–1996

School District	Score % above SAT	4-yr. completion rate	% LEP	% Free and reduced lunch
LAUSD	12.8	63.1	54.1	44.1
Canoga Park	11.5	67.3	36.0	43.9
Chatsworth	25.6	70.0	16.2	19.4
Garfield	5.9	89.6	37.4	70.9
Hollywood	6.7	56.3	75.8	47.1
Jefferson	1.4	48.9	67.7	46.0
Manual Arts	2.1	32.0	66.9	39.8
Monroe	13.9	50.5	49.5	53.9
Narbonne	20.0	—	21.8	38.7
Van Nuys	39.0	69.5	45.7	33.4
Southgate	5.5	76.3	23.7	70.2
Claremont	46.3	92.3	6.8	10.1
Glendora	28.2	91.1	4.6	11.0
Las Virgines	47.1	95.3	3.4	1.4
La Canada	60.9	96.8	4.7	1.7
Manhattan Beach	36.3	95.7	3.5	9.8
South Pasadena	51.5	97.8	4.5	9.8
Torrance	28.1	95.3	15.1	12.0
Walnut Valley	34.2	96.9	7.0	2.9
West Covina	10.2	92.9	15.8	25.5
William S. Hart	26.9	90.2	5.9	7.3

Source: Los Angeles Unified School District, *The County School Districts*; Educational Demographics Unit, State of California, *Language Census Report for California Public Schools*; Los Angeles Unified School District, *School Accountability Reports.*

Hispanic students in the school district. Using the additional information supplied by the free and reduced-price lunch proxy, we can reasonably argue that the Hispanic poverty population in the Los Angeles District is at least associated with the low levels of achievement in the district. In essence, I am arguing that we know that there are individual student-level associations between poverty and performance; we also see aggregate relationships between poverty and performance at the school level. Thus we can conclude that poverty is a critical variable in how students are doing in the Los Angeles region.

Schools and school districts vary considerably. For example, Southgate and Van Nuys[6] have quite different percentages of students scoring above the national SAT level, proportions of LEP students, and levels of poverty. Only 5.5% of Southgate's students score above the national SAT level and 24% are LEP, while Van Nuys has 39% above the national SAT score and 40% LEP students. However, in general, there is a strong association between SAT scores, completion rates, and proportions of LEP students.

Measures of the strength of the relationships, however, as portrayed in Table 5.9, show that there are indeed links between the measures of achievement, poverty, and LEP status. In this group of schools and school districts, the correlation between percentage above the average SAT scores and completion rates is .67; the relationships betwween SAT levels and the percentage of LEP students is –.72. That is, SAT scores decline as the proportion of LEP students rises. Even more critical for this argument is the strong negative association (–.90) between completion rates and LEP students, and between poverty, as measured by the proxy of free and reduced-price lunch and low achievement (–.78). These results reinforce the problem of the emerging link between immigration and achievement. The results would be less troubling if there were numerous employment opportunities in which language and achievement were not crucial, but we have already shown that the opposite is increasingly true.

The plot of completion rates, SAT scores, and LEP ratios for selected Bay Area schools suggests a similar pattern of lower scores and poorer completion rates in those areas that are particularly sensitive to large-scale immigration (Figure 5.18). The schools with large numbers of LEP students are the schools with low SAT scores. It is not a simple one-to-one relationship because some schools with moderate numbers of LEP students perform reasonably well. In general, however, SAT scores are very low in the schools with large numbers of new immigrant children or the children of immigrants. The results reiterate the difference between disadvantaged and advantaged children in terms of success in college.

The evidence from these studies of school districts in two major entry-point regions is almost certainly replicated in other school districts that are centers of immigrant growth. The data strongly support the less sanguine findings of the study conducted by Schoeni, McCarthy, and Vernez (1996). Although some immigrants do catch up to the native-born in income, in general the low incomes have continued over time, especially for Hispanics. The outcome is a continuing cycle of low income and poverty, with consequent effects on children who grow up in such households.

The scale of the problem I have discussed here is greater than suggested by anecdotal evidence of poor performance for individual inner-city schools. The long-term implications are serious for both local communities and the citizen immigrants in those communities. We can no longer assume that the problem will take care of itself, that somehow immigrants will acquire the skills needed to participate in a changing economic milieu, especially in California. Unpalatable as it may be to some constituencies, our society must make a choice between adequately educating the citizen chil-

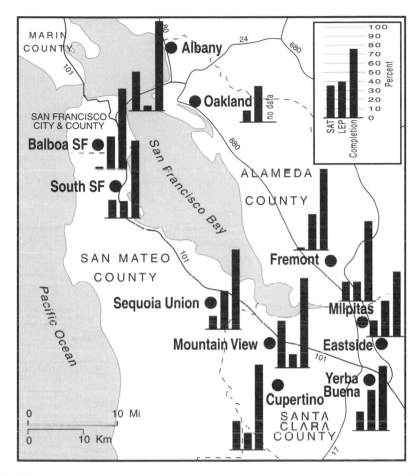

FIGURE 5.18. SAT scores, proportion of LEP students, and completion rates for selected Bay Area high schools, 1994–1995. *Source:* Department of Education, State of California (www.cde.ca.gov).

dren who are already here and who will be born to the increasing numbers of immigrant mothers, or continuing to accept the large-scale flows of poorly educated immigrants, who are likely to overwhelm an already strained education system. At the same time, without additional resources, the current society is merely postponing the serious problems of large-scale failure in the school systems. The concerns expressed by Bean et al. (1994) require the dual decisions of additional resources and a rational immigration policy. Adult education, alternative high schools, and vocational education are critical for saving the resource of the citizen children of the current foreign-born population.

SUMMARY AND OBSERVATIONS

The evidence that many immigrants become homeowners—some groups extraordinarily quickly—is compelling evidence of immigrants' successes in California. Homeownership rates that approach those for the native-born non-Hispanic white population, and that significantly exceed the rates for black households, are important indicators of the ability of immigrants to succeed in the new environment and translate their success into a firm stake in the established order of things. Many immigrants from diverse backgrounds—Canada, Europe, China, Korea, and the Philippines—have become owners. In turn, the corollaries of homeownership are stable neighborhoods, community involvement, and political participation.

At the same time, as our models show, income is a critical dimension of becoming an owner. The substantially poorer immigrants of Mexican origin have greater difficulty in becoming owners, and this fact has longer term implications for asset accumulation. Furthermore, immigrants who arrived in the 1970s, before the very rapid rises in housing price in the late 1970s and early 1980s, entered the housing market at a more propitious time. The very large flows of recent immigrants coincide with high prices and relatively modest increments to the housing stock. Some of these immigrants are concentrated in less adequate housing, rented or owned, in inner-city areas with poor services and high crime. These barrios reinforce other social problems, and counter the success story of the many Asian and Middle Eastern immigrants who have been able to enter the suburban housing market.

The immigrants who have entered the homeowner market are well on their way to economic incorporation and acculturation. In the past, generations of immigrants used the housing market and homeownership to gain an economic foothold in their new country. The evidence for selected groups of immigrants in California shows that this path has not changed. This is the positive side of the story, a side that is integral to assessing whether, and to what extent, the future will bring assimilation or balkanization.

In contrast, the evidence from the analysis of educational gains is mixed. The data show that language skills are being acquired slowly, that the number of LEP students is growing, and that at least some immigrant groups are finding it difficult to gain the skills that will lead to greater economic benefits. Not all immigrant groups are having the same transition difficulty. Among some groups, especially of the Asian and Middle Eastern immigrants, educational achievement levels are high and human capital is increasing.

In general, the Latino population is most disadvantaged and is suffering most severely from the lack of human capital. The immigrant Hispanic population, although perhaps better off in California than in rural villages in Mexico, is trapped in a cycle of inadequate education and low-paying jobs. Also, the current school systems and the reliance on bilingual education are apparently not overcoming the problems of educating an increasing segment of the citizen population of California. Although money alone may not be the entire solution, there is no question that California must move from the bottom tier of state school funding to the top, if the generations of new immigrant students are to be incorporated into the postindustrial society.

NOTES

1. Recent reports for the United States as a whole show that the rate of homeownership for the foreign-born is virtually identical with the rate for the native-born (Callis, 1997).
2. The combination of high population densities, crowded housing, and poverty characterizes inner-city barrios and ghettos (Jargowsky, 1996). I have already noted that residential crowding has returned to high levels in Southern California, especially for Hispanic immigrants (Myers and Lee, 1995). Inner-city barrios and ghettos are likely to be crowded and to have very high levels of poverty and other social problems. The data show that high-poverty neighborhoods have increased in U.S. metropolitan regions (Jargowsky, 1996) .
3. "Very crowded" is defined as the percentage of units with more than 1.5 persons per room.
4. The data are not available for California separately.
5. The debate about bilingual education has recently reached the courts in California. In August 1997 a Sacramento County judge halted the Orange Unified School District's new English immersion program. The judge stated that he was not satisfied that the school district had safeguarded the rights of non-English-speaking children. That decision and a ballot measure that would require teaching in English only are symptoms of the tensions related to large-scale increases in the foreign-born population. Overall, the decisions reflect once again the willingness of groups to use the legal process to protect, or challenge, the stakeholders in the changing immigrant context of California. The idea that educators should try to make the best decision seems to have been lost in the process, and the divisions among stakeholders are exacerbated rather than ameliorated. In June 1998 61% of Californians voted in favor of Proposition 227 that will require all public school instruction to be in English.
6. Van Nuys has a math-science magnet program.

CHAPTER 6

ASSIMILATION VERSUS SEPARATION:

IMMIGRANTS' FUTURE TRAJECTORIES

The very large recent flows have revived old antipathies to new immigrants, and have led to loud calls for radical changes in our immigration policy and for increased controls on cross-border migrations from Mexico. In previous chapters I argued and documented that the composition of immigration flows is different from those at the turn of the century, that current immigrants are unlike those who arrived earlier, and that the new immigrants are entering a changing economy. Many are doing less well than earlier entrants. Moreover, changing sensitivities to language and culture have emphasized preservation of the new immigrants' language, culture, and identity, which is slowing their assimilation.

This emphasis, plus the very size of the immigrant flows, has created concerns about the separateness and even the isolation of the new immigrants. Will the emphasis on pluralism and multiculturalism rather than on the "melting pot" change the United State's basic commitment to a "blended society" (Schlesinger, 1992; Clark and Morrison, 1995)? What does this mean for California's future?

A related question is whether California has the capacity—political, economic, social, and cultural—to absorb the many immigrants currently entering without prompting serious negative effects on social institutions, especially in those local communities to which immigrants flow disproportionately. Must our society reduce the current levels of admittance to enable it to absorb the recent immigrants? Advocates of a slowdown point both to the economic and social costs of large-scale immigration and to a concern for the future of the American way of life (Brimelow, 1994). Certainly, the appeals to anti-immigrant sentiment are not based on dollar costs alone (Griego, 1994; Lopez, 1996). Lopez, for example, traces much of the negative reaction to recent immigrants to linguistic differences between "them" and "us" and their slow progress in adopting English.

In addition, the service sector, with whom the white non-Hispanic majority comes into contact most frequently, is dominated increasingly by recent immigrants. This contact has exposed an increasing separateness within Californian society. Increasingly the workforce in California speaks Spanish: the restaurant waiter takes orders in English and conveys them to the cook in Spanish; the Mexican gardener hires a Spanish-speaking helper to rake leaves. None of these jobs affords real opportunity for advancement—of the kind that enabled immigrants in earlier eras to advance economically and merge into the mainstream. Is this change temporary or permanent? In this context it is important to consider the data on changes in language use, levels of naturalization, and the extent of intermarriage to evaluate the patterns of assimilation. Is there a new pattern of assimilation? Is there no assimilation? Or do these concrete experiences conceal a bifurcated pattern of socioeconomic success and assimilation for some immigrants, and separation, deprivation, and poverty for others?

THE RELEVANCE OF THE "MELTING POT" METAPHOR

The metaphor of the "melting pot," in which new immigrants eventually are "Americanized," has always concealed a vastly more complex reality. The "Little Italies" and German and Polish neighborhoods in almost every eastern and midwestern city testify to the continued vitality and viability of ethnic groups long after the immigration flows from their countries of origin ended in the second decade after the turn of the century. Yet the children of these Italian, Polish, German, and other European immigrants have left behind both the language of the "old country" and their old ethnic neighborhoods. This fact suggests that even if the "melting pot" is an overstatement, the idea had elements of truth. The enduring notion that second- and third-generation Americans could become whatever they wanted to be, that they were not bound either socially or spatially by the concepts and strictures of the "old country," is central to the idea of a new, "integrated" society. Survey results support this notion for Hispanics (NALEO, 1989; Lopez, 1996). Moreover, Asians' levels of entry to California universities offer a different kind of testimony to the continuing power of the notion of moving upward in our society, and of the concomitant likelihood of assimilation (Abelmann and Lie, 1994).

In the past, the "melting pot" metaphor invited faith in the notion that a new "American" society was continually re-created out of immigrant influx. Ideally, the work of the melting pot, the process of cultural assimilation, took place over time. Integration into the host society might span several generations, so that the third- and fourth-generation descendants of earlier waves of immigrants only now are fully blended into the larger society. Much of this process is related to increasing years of education and transitions from blue collar to white-collar and professional occupations, and to social change across generations.

The earlier immigrants often came from rural areas and were mostly unskilled, but their children went to U.S. grade schools and high schools and moved into skilled employment as electricians, plumbers, and autoworkers. Many of the children of the second generation entered college to become the teachers, lawyers, managers, and professionals in an increasingly technologically oriented society. This process took place for countless earlier immigrants from Europe, but not for all immigrants; some were left behind to scratch out a precarious living in inner-city ghettos. Then, as now, the economic context was important, and skills were the key to successful immigration.

Cultural assimilation (or "acculturation," as it is sometimes called) is the first in the series of stages in the assimilation process. Gordon (1964) traced out the classic process from initial migration to language acquisition to intermarriage. This latter step is the ultimate form of assimilation because the children of intermarriage (e.g., between an Irish American and an Italian American) are a melding of cultures and traditions. Cultural assimilation is the adoption of the mainstream language and participation in the mainstream educational process; these steps are important because language and education lead to occupational mobility and higher incomes. These in turn can lead to social and marital assimilation.

For most Americans the process is largely complete, even though earlier waves of immigrants retain strong linguistic and cultural affinities with the "old country"; think, for instance, of the Italians in New York. At the same time, few if any socioeconomic differences exist between non-Hispanic white ethnic groups. High levels of intermarriage across ethnic lines have blurred any significant differences, such that the former gulf between northern and southern Europeans has vanished (Lieberson and Waters, 1988). White groups are more like than unalike each other on a wide range of economic, occupational, and educational measures (Lieberson and Waters, 1988).

Cultural assimilation and socioeconomic progress are intertwined not only with the times but with particular places; I focus on the latter in this book. When immigrants first arrive, they cluster in central locations dominated by members of their own ethnic group, a tendency that is no different in 1998 than it was in 1908, although it appears to be more true for the poorest immigrant groups, while those with some resources locate near their friends and relatives in more dispersed locations (Allen and Turner,

1996). Such centrally located enclaves provided then, and still provide today, a support system that includes housing, a sense of community, and jobs. The concentrations of the foreign-born in Los Angeles and San Francisco, which I examined in earlier chapters, are the spatial expression of the process of coming to a new society and establishing an initial foothold in the new cities.

Logic suggests that as immigrants improve their situation and their socioeconomic status, they will also undergo residential assimilation. Indeed, indices of dissimilarity decline with increasing economic and educational levels for both Hispanics and Asians (Clark and Ware, 1997; Denton and Massey, 1988). As a corollary we might expect to see Hispanic and Asian professionals moving to the suburbs, where the overall socioeconomic status is higher than in the cities. If spatial assimilation also occurs, then we are witnessing the classic pattern followed by earlier European migrants as their socioeconomic status improved. Massey and Denton (1987, p. 818) firmly support the idea that residential assimilation accompanies economic gains, and they conclude that Asians as well as Hispanics are following processes of integration and spatial assimilation similar to those of earlier European immigrant groups (Denton and Massey, 1988). These findings are consistent with much earlier work showing that urban deconcentration was associated with decreasing segregation of ethnic groups (Duncan and Lieberson, 1959; Lieberson, 1963).

The notions of immigrants' aspirations, educational and professional achievement, and dreams of suburbanization occur frequently in discussions of immigration trajectories. Duran and Weffer (1992) emphasize the role of high school progress in immigrants' aspirations, while Abelmann and Lie (1995) emphasize the suburban dreams of Koreans in Los Angeles. Although Kao and Tienda (1995) argue that there are multiple paths to assimilation, they conclude that increased academic achievement by native-born children of immigrants is the real key to success, and that immigrant parents play an important role in stimulating their children's achievement. The authors further suggest that data that shows that third generation immigrant children do less well than first- and second-generation children is not evidence that assimilation is not working. In fact, they argue that such a dropoff actually proves the success of assimilation: as distance from the first-generation immigrant household increases, achievement levels are likely to be more similar for all children.

Assimilation has never proceeded simply or in a straight line. Insofar as immigrant groups "melted" into U.S. society, the process was awkward and incomplete. The process of adaptation was painful, and ethnic resilience in the face of pressure to conform to U.S. ways was a way of protecting the group and its members. Even though it was a slow and imperfect process, the "white" immigrants did eventually merge and blend. The current immigration process, however, may not play out in quite the same way.

Some observers reject the necessity of any assimilation. Glazer (1993) points to evidence that new forms of cultural adaptation are already occurring in American metropolitan areas. He suggests that because assimilation worked only with Europeans, and much less perfectly, if at all, with blacks or Asians, it is questionable whether any nonwhite racial and ethnic groups ever succeeded in truly assimilating. Glazer cites widespread support for bilingual education and foreign language rights, and the acceptance of protected status, as indications of the breakdowns of assimilationist thinking in U.S. society. But even if assimilation has not worked for African Americans, there is evidence that it is working (though slowly) for Asians, the other most distinctive racial group in the United States. How this situation will evolve in the long run is a critical aspect of future American society.

The idea of ongoing changes in the assimilation process is taken a step further by Portes and Zhou (1994), who argue that when economic opportunities are much less favorable, groups that find it harder to assimilate because of racial rather than ethnic differences will remain separate. Thus an increase in adversarial positions by immigrant groups will create different approaches to assimilation and acculturation. These new forms include downward acculturation, in which immigrants refuse to follow the path of melting pot assimilation and choose to emphasize their differences and distinctiveness as a means of achieving economic success. This change is also a response to the "hourglass economy," which has replaced the old ladder of opportunity for the new ethnic groups of the late 20th century (Fainstein, Gordon, and Harloe, 1992).

Even though surveys show general support for preserving immigrants' customs and cultural identity, U.S. society at large still expects some form of adaptation to the larger "American" society. Moreover, insofar as some groups choose to emphasize their differences, they raise the possibility of conflict rather than cooperation. For example, the waving of Mexico's national flag to celebrate Mexican Independence Day can be interpreted either as a well-intentioned celebration of Mexico's rich history or as a statement by a nation within a nation, and hence a sign of balkanization.

A conflict exists between research that emphasizes findings that the new immigrant children are intent on not rejecting the "old world" and see no need to do so (Portes and Rumbaut, 1996) and the continuing stream of survey data emphasizing the desire for integration. In a recent Gallup national poll, almost 60% of the respondents felt that it was better to blend in than to remain separate (Saad, 1995). The general survey findings are supported by the strong aspirations for social mobility through education, exactly the same process that helped earlier generations to integrate. Earlier research in Southern California, focusing on Hispanics only, provided some evidence of upward and outward mobility (Clark and Mueller, 1988). That study found that Hispanics with higher education and occupational levels were moving to suburban locations, though their spatial assimilation was less clear. To what extent are those conclusions still valid? Does recent evidence support the idea of less rather than more assimilation?

MEASURING ASSIMILATION

Assimilation, as we now recognize, is multidimensional. The old metaphors no longer fit the realities visible today in California. How should we think about this concept? How can it be measured? Assimilation is a process that involves succeeding generations that takes place slowly over decades. Unfortunately, census data does not provide sufficient detail on the children of the children of immigrants. We have to rely on imprecise barometers such as data on language acquisition and use, on the rate of naturalization, and on the levels of intermarriage. Such data, if used cautiously, however, can be used to assess the path of assimilation. Intermarriage, of course, is regarded as the ultimate sign of assimilation because the children of intermarriage are no longer of one ethnic ancestry or another but a mix. Like so much of the discussion of assimilation, the focus is on the subjective assessment of whether the glass is half full ("Assimilation will work but it may take a long time"), or half empty ("The new immigration will create patterns of separation and ethnic identity which will slow if not prevent the "melting" into a new society").

Much of the vigorous debate concerns a larger issue: How is American society to be defined and what does it mean to be an "American"? Many people wonder whether the celebration of diversity is creating a rift in our social fabric (Schlesinger, 1993). Even those who support the pluralistic approach (the incorporation of many perspectives on race and ethnicity) are concerned about large-scale immigration leading to social stresses and even to national division (Ravitch, 1990). This group believes that U.S. society should promote our common humanity, not our group ties. Schlesinger's (1993) worry about a balkanized future is the most extreme of these positions, which examine how dividing America into fixed ethnicities will influence the future of the Union. Certainly, much of this concern is related to uneasiness felt by the (still) white majority about the future structure of U.S. society. America's rapidly changing racial composition underlies the worry about multiculturalism and whether or not America will become a Tower of Babel (Hirsch, 1987).

The evidence of the past indicates that the United States welcomed large numbers of poorly educated and low-skilled people from Europe and eventually turned them into productive and successful American citizens. At the same time, however, it is clear that this process failed with blacks, and that it worked only partially for earlier generations of Chinese or Japanese. The marginality of those groups has stimulated the notions of cultural pluralism as a counter to the simple ideas of the melting pot. The notions of strength in diversity, identifying with one's past, and recognizing the advantages of a shared heritage can be viewed as complements to becoming an American. This pluralism is central to the emergence of multiculturalism and to an emphasis on language preservation.

Language Assimilation and Fluency in English

Language is the first and perhaps the most important step on the way to assimilation. Without fluency in the language of one's new home, one feels like an outsider—not one of the group, unable to participate in the social processes of a new society (Hoffman, 1989). Language, after all, is the means of communication and the key to education. The increase in use of non-English languages certainly contributes to native Californians' fear that their state is becoming fragmented into ethnic enclaves, each with its own dialect. What is the structure of language use in California? How is it changing? Does the future hold the concoction of blended languages depicted in the film *Blade Runner* (also mentioned by Lopez [1996] in an insightful essay on language), or will language be the first step to assimilation, as it has been in the past? The debate over English language use and fluency raises the essential dilemma for immigrants undergoing the assimilation process: maintaining respect for their own heritage (the language of the country they left behind) versus equipping the new generation to prosper in a world where English proficiency is demanded for all but the lowest level jobs.

Not surprisingly, with so many new immigrants, an increasing number of individuals in California speak a language other than English at home. The specific language is closely related to the individual's birthplace (Table 6.1). Spanish is the most commonly used "other" language; it is the household language of almost 5.5 million people in California. The Asian languages—Chinese, Korean, Tagalog (from the Philippines), and Japanese—as a group are the second-most commonly used "other" language; they are spoken by some 1.5 million. The predominance of Spanish emphasizes the sheer size of the recent immigration.

Because the native-born have frequent contact with people in service jobs (e.g., gardeners, gas station attendants, waiters), and because so many of the people working in service jobs are immigrants who speak English

TABLE 6.1. Languages spoken at home in California by more than 50,000 persons

Language	Number
English only	18,764,213
Spanish	5,478,712
Chinese	575,447
Tagalog	464,644
Vietnamese	233,074
Korean	215,845
German	165,962
Japanese	147,451
French	132,657
Italian	111,133
Portuguese	78,232
Arabic	73,738
Hmong-Khmer	59,622

Source: U.S. Bureau of the Census, 1993.

haltingly (if at all) and therefore typically employ their native language (most often Spanish), the native-born sometimes feel as if they are the ones speaking the "foreign" language. Spanish is not confined to the barrio: today, it is the language of construction workers, of repairmen, of janitors, and increasingly of the people who serve in the department stores, supermarkets, and fast food outlets. California's laborers, busboys, and gardeners speak Spanish, and the ubiquitous mobile food trucks that roam the streets of Los Angeles are moving centers of the Spanish language in all parts of the city.

California is quite different from the national pattern: the state contains more than a quarter of all U.S. households that use a language other than English. Thus it is easy to understand why people perceive a divided and separate society.

The fact that nearly one-third of the California population speak a language other than English in their homes has created a sense of language division (Table 6.2). One obvious division exists between the native-born English speakers and the immigrants. Another division exists between Hispanic immigrants who speak Spanish at home and the smaller number of Asians who speak one of the Asian languages.

In the native-born population overall, only 14% speak a language other than English at home. Among native-born Asians and Hispanics, however, much higher percentages speak a language other than English at home. That proportion reaches 62.5% for Southern California native-

TABLE 6.2. Population speaking a language other than English at home, by nativity and ethnicity, 1990

	Total (%)	U.S.-born (%)	Immigrants before 1980 (%)	Immigrants 1980–1990 (%)
California	31.4	14.0	78.6	92.2
Southern California				
All residents	39.3	17.3	82.2	93.5
Asians and Pacific Islanders	80.3	40.7	88.9	95.7
Hispanics	81.5	62.5	95.2	96.2
Other	10.0	4.7	48.6	75.5
Bay Area				
All residents	29.7	11.7	76.4	91.0
Asians and Pacific Islanders	67.3	41.5	89.1	95.3
Hispanics	78.5	46.1	92.3	95.6
Other	9.2	4.7	47.9	66.4

Source: U.S. Bureau of the Census, 1992.

born persons speaking Spanish at home. The proportion is 46.1% in Northern California. For native-born Asians and members of other ethnic groups, the percentages are significantly lower (Table 6.2). For immigrants, as we would expect, the percentages are much higher. The proportions fall below 50% only for "other" immigrants, including Canadians and other native speakers of English. Among the most recent immigrants, more than 90% of the population speaks a language other than English at home. It is useful to remember that apart from the very high level of Spanish speaking at home by the native born population, these figures do not differ from what we know to be true of immigrants who arrived early in the century.

Most of the native-born ethnic population has already developed English language skills. Large numbers of native-born Asians and other second-generation immigrants either speak only English at home or speak English well (Table 6.3). The differences between those with less than a high school education and those with college degrees are not large in this context (Table 6.3). However, the percentages of Hispanics who speak only English at home or who speak English well are much lower than the percentages for other groups. Native-born Hispanics show less language acculturation than do other immigrants. Educational level marks a significant difference in English proficiency for the foreign-born population: Asian and Hispanic foreign-born immigrants are three to four times more likely to speak English well if they have a college degree than if they have only a high school education (Table 6.3). Finally, we see some differences

TABLE 6.3. Language use and English ability in California by education and nativity, 1990* (Asians, Hispanics, and others aged 25–44)

	Speak only English at home* (%)				Speak English very well* (%)			
	All	Asian	Hispanic	Other	All	Asian	Hispanic	Other
Native-born								
Less than HS education	79.8	75.7	39.6	96.3	70.6	45.1	70.5	73.9
High school graduate	89.4	83.3	53.5	96.8	81.4	71.1	82.3	79.4
Some college	91.3	82.1	56.0	96.1	83.8	79.1	85.4	82.0
Degree plus	92.5	80.2	57.7	95.1	85.3	82.6	88.1	84.4
Foreign-born								
Less than HS education	5.0	3.5	3.8	29.3	16.5	15.8	15.9	32.9
High school graduate	14.3	7.0	4.9	46.6	38.8	33.3	38.1	54.2
Some college	20.9	9.1	6.8	54.6	51.7	44.4	51.9	70.6
Degree plus	20.8	8.6	9.9	45.5	63.8	60.9	57.2	76.7

*Of those persons who speak another language in addition to English.
Source: U.S. Bureau of the Census, 1992.

between Northern and Southern California. The gap in English proficiency between native-born and foreign-born is generally smaller in Northern California; immigrants in that area are more likely to use English in the home and to speak English well, but the differences between foreign-born and native-born persist (Table 6.4).

Clearly, more languages are spoken by more people today in California than in the past. Clearly, too, more people are speaking different languages. But the real issue is whether the immigrants' use of foreign languages will persist. We hear countless stories of United States–born Japanese, Korean, and Chinese who are expected to speak the language when they travel to their ancestral countries. After all, they look Japanese, Korean, or Chinese to the people who live in their ancestral homelands. In fact, however, they typically have limited (if any) skill in the ancestral language because it has been lost in the transition to the new society. This loss is due to assimilation. Thus the question to be addressed is the extent to which immigrants are passing on their native language to their descendants.

Adult immigrants continue to use their mother tongue as their primary language, especially in the home. Thus the second generation, growing up in a home where their parents' native language is regularly spoken, re-

TABLE 6.4. Language use and English ability in Northern and Southern California by education and nativity in 1990 (Asians, Hispanics, and others aged 25–44)

	Speak only English at home* (%)				Speak English very well* (%)			
	All	Asian	Hispanic	Other	All	Asian	Hispanic	Other
Northern California								
Foreign-born								
Less than HS education	5.9	3.1	4.0	31.2	17.3	12.2	18.8	32.9
High school graduate	13.8	6.3	5.7	45.7	38.4	31.8	40.6	59.9
Some college	19.3	8.5	7.3	52.0	51.1	43.2	54.6	69.9
Degree plus	20.8	8.5	12.7	48.8	65.8	62.6	65.1	78.1
Native-born								
Less than HS education	83.3	74.1	49.8	95.8	72.6	38.4	72.8	79.2
High school graduate	90.6	77.6	61.7	96.7	80.4	70.5	81.0	80.8
Some college	92.4	75.2	65.3	96.3	83.9	76.4	86.9	83.0
Degree plus	92.5	75.7	62.5	95.0	85.0	80.6	88.5	85.3
Southern California								
Foreign-born								
Less than HS education	4.5	3.2	3.5	26.5	15.4	16.0	14.7	31.3
High school graduate	11.5	5.8	4.3	39.0	35.8	28.8	35.4	49.7
Some college	17.6	7.2	6.1	49.1	49.3	41.0	49.4	68.4
Degree plus	17.4	7.5	8.4	38.6	61.6	58.7	52.7	76.0
Native-born								
Less than HS education	77.6	75.3	40.5	95.8	69.7	52.0	69.4	73.2
High school graduate	88.1	85.1	52.9	96.7	81.6	67.8	81.7	81.9
Some college	90.2	85.7	55.0	95.8	84.1	83.4	85.7	81.6
Degree plus	91.9	81.6	56.0	94.8	86.5	87.5	87.8	85.6

*Of those persons who speak another language in addition to English.
Source: U.S. Bureau of the Census, 1992.

tain the language. But the second generation goes to U.S. schools and learns English. Thus outside the home, they use English as their first language. The third generation grows up in homes where English is the norm; they do not learn the language of their grandparents. This model of language shift, which was formulated originally for European migrants (Fishman, 1972) also applies to Asian immigrants and even to Spanish speakers

(Lopez, 1978, 1982, 1996). But how well does it describe a situation that is quite different from the European immigrants' context? Spanish speakers are not isolated from the hearth language, and their numbers are sufficient to allow them to continue to speak, work, and live in a Spanish-speaking world.

As a quick response to the closeness of Mexico and the very large numbers of Latino immigrants, one might say that little if any shift to English will occur. Certainly the data in Tables 6.3 and 6.4 suggest that Hispanics are adopting English more slowly than Asians. If this is true, what do the data tell us about language use over the generations? To answer this question we need data on Hispanic parents, their children, and their children's children, but such data are limited. We do have some information, however, on language use by year of arrival, and some selected surveys can be used to enrich the census results.

Language use differs with time of arrival, and the data also reemphasize interethnic differences (Table 6.5). Immigrants who arrived before 1960 are very likely to speak English well. Almost 90% of Middle Eastern immigrants who arrived before 1960 now speak English fluently. The proportion is lower for Mexican and Central American immigrants. The most significant findings reported in Table 6.5, however, are that for all immigrant groups except Latinos, only the most recent immigrants do not speak English well, and that all groups make rapid strides in developing English language skills.

The fact that Latinos have lower levels of English language skills and are much less likely to speak English at home appears to validate the notions of proximity and size that I discussed in the preceding paragraphs. At the same time, data from Southern California on the percentage of the Hispanic group who speak only English at home suggest that the generational model *is* working for Southern California (see in Table 6.6). The shift is

TABLE 6.5. Language skill by ethnicity and year of arrival

	% speaking English well <1960	1960–1969	1970–1979	1980–1990
All	79.2	75.7	70.1	53.3
Mexico/Central America	68.8	65.6	59.6	39.1
S.E. Asia	*	86.3	76.1	53.5
Other Asia	76.6	86.8	82.0	66.7
Middle East	89.3	92.2	90.3	74.7

*Insufficient data.
Source: U.S. Bureau of the Census, 1992.

TABLE 6.6. Percentage speaking only English at home, by ethnicity and generation, 1989 (selected ethnic groups, aged 25–44), in Southern California

	All Hispanics (%)	Mexicans (%)	All Asians (%)
First generation	3.0	2.0	15.0
1.5 generation	13.0	6.0	—
Second generation	28.0	28.0	75.0
Native of native-born	57.0	57.0	92.0

Source: Lopez, 1996. © 1996 Russell Sage Foundation. Reprinted by permission.

dramatic: while only 3% of first-generation Hispanics speak English as their primary language at home, almost 30% of second-generation of Hispanics speak only English. The percentages for Asians are quite remarkable: 75% speak only English at home by the second generation.

It is impossible not to conclude that assimilation is occurring and that the children of immigrants are acculturating to the English language. That the shift is so much faster for the Asian population supports the hypotheses of size and separation suggested above. The analysis of language thus confirms what has been a consistent theme throughout the book: we cannot simply speak of "immigrants" because they differ fundamentally in origin and cultural background. Also, even within the divisions I have used here, not all Asian immigrants have strong language skills and not all Mexican immigrants are LEP students. For the immigrant groups overall, language acquisition indicates the potential for immigrants' absorption and adaptation. In some groups, however, the process will be slower and more difficult, with the attendant problems in acquiring other skills and fully entering the labor market.

Assimilation via Naturalization and Citizenship

Naturalization is a useful gauge of assimilation because it reflects progress in, and familiarity with, English and because it signifies a process of separation from the country of origin. One might reasonably expect that those who do not speak English well are much less likely to apply for citizenship than those who do speak English well. One might also logically suppose that those with strong ties to their country of origin are likely to resist allegiance to a new society. Naturalization is related to length of residence, too: those who have been in this country the longest are more likely to be naturalized (Table 6.7). Immigrant groups naturalize at very different

TABLE 6.7. Rates of naturalization for selected nationalities

Ethnic background	Year of arrival			
	<1970	1970–1979	1980–1989	1990>
Canada	.88	.16	.19	0
Europe	.80	.69	.48	.01
Asia	.78	.71	.40	.04
Mexico/Central America	.34	.17	.05	.01

Source: U.S. Bureau of the Census, 1992.

rates. In general, Canadians and Mexicans have been the slowest to become citizens (Table 6.7), presumably because the ease of maintaining ties with Canada and Mexico stands in the way of forming a new allegiance to the United States.

It is expected that recent immigrants would not be citizens; migrants who are not already married to U.S. citizens must wait 5 years to apply for citizenship. Even so, for arrivals before 1980, Asians are almost twice as likely as Hispanics to be naturalized. Early arrivals are simply much more likely to be citizens. Also the groups that are further removed from their origins are more likely to be citizens because they are unlikely to return permanently to their original homes. For some Asians, naturalization also represents a commitment to the new society.

The county-level rates of naturalization in California reflect both the timing of immigration and the composition of the immigrant population in the counties of destination. Once again we see the geographic bifurcation of timing and composition (Figure 6.1). In the five northern counties in the Bay Area, more than 50% of all immigrants who arrived between 1965 and 1979 are now citizens. In contrast, in Southern California, the percentages are in the 30s. These different percentages emphasize the very different trajectories of immigrants in these separate locations.

The map of naturalization rates by county (Figure 6.2) further emphasizes the distinction between Northern and Southern California. It also highlights the very low rates of naturalization in many of the agricultural counties. Among immigrants who arrived before 1980, naturalization rates are as low as 10% in most of the Central Valley communities.

The county data mask very high rates of noncitizenship in cities and communities. In Los Angeles County, for example, in Huntington Park, 49% of the population were not citizens. In the cities of El Monte, South Gate, and Santa Ana, nearly 40% were noncitizens. In the cities of Alhambra, Monterey Park, Glendale, Lynwood, and Baldwin Park, one-

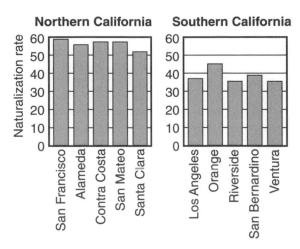

FIGURE 6.1. Naturalization rates of recent immigrants (1965–1979) in Northern and Southern California. *Source:* U.S. Bureau of the Census, 1993.

third of the population were not citizens. All of these communities contained immigrant populations that were heavily Hispanic and largely Mexican.

Do these figures mean that Asians are more likely to be assimilated? The evidence, at least for the structural changes of learning English and becoming citizens, is strongly positive. Does this suggest that a Mexican nation within a nation is emerging? Not necessarily, although Rodriguez's (1996) findings emphasize Hispanic immigrants' links to Mexico and the role of cheap travel and low-cost telephone service in preserving a cross-cultural link and slowing any tendency to assimilate to a new, blended society.

Assimilation and Intermarriage

Intermarriage is critical in the assimilation process, as it is the final step to merging two different cultures and creating a new blended society. The child of white and Asian parents or the child of Asian and Hispanic parents no longer has a single ancestry but represents a mixing of races and ethnic backgrounds. The year 2000 census, for the first time, will acknowledge such multiple racial and ethnic memberships. High rates of intermarriage are an indication of brisk social interaction, but because social contacts tend to take place in a social and spatial context that emphasizes contact

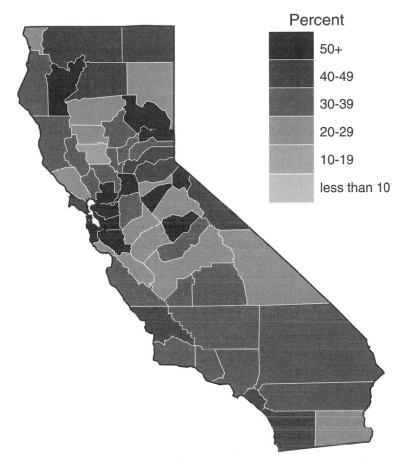

FIGURE 6.2. Naturalization rates of recent immigrants (1965–1979) by county.
Source: U.S. Bureau of the Census, 1993.

with "like" people, marriage within groups is most likely. Thus marriage across ethnic and racial lines is the measure of true melting and of assimilation. It is also probable, because marriage across widely different groups is much less likely, that the process of intermarriage will be slower today than it was for earlier, basically European groups (Root, 1992).

The likelihood of marrying within or outside an ethnic group depends to a certain extent on the size of the group. A small ethnic group with a small number of eligible partners will inspire a search for a marriage partner outside the group. In contrast, very large ethnic groups, such as the Mexican population, will contain a large number of prospective spouses.

At the same time, however, the increased number of different ethnic groups in California and the increase in contact between them is likely to heighten the propensity to marry outside one's own ethnic group. This will be more true for younger people than for older people because many of the younger members of immigrant groups married after they migrated to California. It is also likely that younger, more recent arrivals are more receptive socially to other ethnic groups; however, there is no specific research on this topic.

Overall, rates of in-marriage are still high, though some groups have quite low rates of in-marriage and we see a clear distinction by age (Table 6.8). Whites sometime married outside their ancestral ethnic group, but in the past the spouse was also white. Research also suggests that ethnic identity is still important even for those individuals who do marry outside their group (Spickard, 1989). Because the dominant population is white, it is the population for which intermarriage would be the strongest indicator of true blending and thus of assimilation. The data show that about 11% of white women aged 25–44 now are marrying outside (Table 6.9).

Intermarriage for native-born white women is highest with the Mexican-origin immigrant population and the "other" immigrant population; the latter group, of course, includes European and Canadian spouses, men who are likely to be of the same ethnicity as the women themselves (Table 6.9). The highest rates of intermarriage occur between groups that are alike in culture and society; thus we do not find very high intermarriage rates between whites and most Asian groups. We expect and observe higher levels of intermarriage between whites and members of Hispanic-origin populations.

Other research on Southern California provides further details on par-

TABLE 6.8. Rates of in-marriage in California, 1990

	% Married women in-married	
	Age 25–34	Age 55–64
White	89.3	96.2
Mexico	79.4	84.4
Central America	68.6	87.3
China	68.3	58.0
Japan	40.6	58.0
Korea	78.8	84.5
Philippines	57.9	83.3
Vietnam	88.1	*
India	84.1	*

*Insufficient data.
Source: U.S. Bureau of the Census, 1992.

TABLE 6.9. Intermarriage rates of non-Hispanic white women in California, 1990

	Age 25–34	Age 55–64
White in-marriage	89.38	96.52
Chinese	.30	.06
Philippines	.36	.09
Japan	.44	.06
Korea	.05	.01
S.E. Asia	.06	.00
Cuba	.15	.02
Central America	.26	.05
Mexico	4.90	1.23
Other	3.44	1.34

Source: U.S. Bureau of the Census, 1992.

ticular groups (Cheng and Yang, 1996; Bozorgmehr, Der-Martirosian, and Sabugh, 1996; Allen and Turner, 1997). The Middle Eastern and Southeast Asian ethnic groups are the least likely to marry outside their own group. The rates of intermarriage among most Asian groups (excepting Japanese and Thais) are low, even though to outsiders these groups may seem culturally similar. Central Americans (marrying other Hispanics), Filipinos, Japanese, Thais, and Puerto Ricans are most likely to intermarry. Central Americans' intermarriage with other Hispanics will not change the fundamental cultural connections of the Hispanic population, and indeed will likely slow assimilation. Cubans and Puerto Ricans are most likely to marry outside their own group, a reflection of the relatively small size of their groups. This fact also indicates that they are likely to begin the assimilation process more rapidly. Almost one-fifth of Japanese, Thais, and Puerto Ricans marry non-Hispanic whites. Quite different patterns exist for Middle Eastern groups, such as the Iranians and Armenians. The very low rate of intermarriage for Middle Eastern groups is due to religious norms. Intermarriage between Muslims and non-Muslims is generally forbidden in Middle Eastern countries (Waldinger and Bozorgmehr, 1996, p. 368).

Intermarriage in the Mexican population provides an important context for judging the process of assimilation. Allen and Turner (1997) report a change in the trend of Mexican intermarriage. In 1963, before the initiation of the recent very-large-scale migration, 25% of all people of Mexican origin were marrying non-Mexicans. In 1990, only 14% of all people of Mexican origin did so. The explanation is obvious. Today, the very large levels of Mexican immigration provide large numbers of marriage partners of a similar culture and background. The Mexican concentrations, in particular, have increased with increased in-group marriage. This development has potential effects on socioeconomic assimilation, social integration, and residential segregation.

Socioeconomic Assimilation

Language affects the ability to learn, and the achievement of higher levels of education in turn influences economic status. We have already seen that new waves of immigrants are doing less well than earlier waves. Such a downward trajectory raises questions about how, and whether, the new immigrants will integrate successfully. But what is the history of the past migrants, immigrants who arrived two decades ago and the native-born children of immigrants? Here again the evidence seems to support a mildly positive view of the power of assimilation and integration. The data from a study of Hispanic households suggest that American-born Latinos, precisely those who should be assimilating today, in fact are joining California's middle class. By 1990 more than half of all American-born Latinos in the Los Angeles region were "middle class," that is, had a household income of $35,000, owned their own home, or both (Rodriguez, 1996). The household data tell a more optimistic story than the focus on individual immigrants. On the basis of the optimistic findings by Rodriguez (1996), the *Economist* (1996, 14 December, pp. 28–29) trumpeted Latinos as "the next Italians." With some differences, eventually they may well parallel those earlier arrivals.

I replicated the Rodriguez study by defining the middle class simply as households with earnings of more than $35,000 a year; I ignored home ownership and the combination of income and home ownership as defining features of "middle class." The Latino middle class, measured by household income, constitutes a significant proportion of all 1990 Latino households. Although my analysis based on income rather than income and home ownership is slightly less optimistic than Rodriguez's, I too conclude that a Latino middle class is slowly emerging. In 1990, 46% of native-born Hispanic households earned more than $35,000 (Table 6.10). This proportion reflects a real increase from the 1980 figure of 34%. For the native-born and for every interval of arrival, the proportion of Hispanics in the middle class increased significantly. Arrivals between 1960 and 1964 were close to 50% middle class. This proportion declines to about a quarter of all households for the latest arrivals, exactly as an assimilation argument would suggest (Table 6.10). Even among the households who arrived between 1975 and 1980, 13% had incomes over $22,040, the 1980 equivalent of the 1990 measure of the middle class ($35,000). In 1990,

TABLE 6.10. Percentage of middle-class Hispanic households in 1980 and 1990*

	Middle class in 1980		Middle class in 1990	
	All Hispanic	Mexican	All Hispanic	Mexican
Native-born	34	33	46	46
Arrived < 1960	30	28	40	39
1960–1964	31	26	49	46
1965–1969	26	23	41	39
1970–1974	18	17	35	34
1975–1979	13	12	30	29
1980>	—	—	25	24

*Defined as incomes of $22,040 in 1980 and $35,000 in 1990.
Source: U.S. Bureau of the Census, 1983, 1992.

30% of the same group belonged to the middle-class; this was a significant change in just a decade.

In 1990, approximately 342,000 native-born Hispanic households qualified as "middle class" under this definition, an increase of more than 60% from the 211,020 households in 1980. In 1980, 132,340 foreign-born Hispanic households had middle-class incomes. By 1990, the number of middle-class foreign-born Hispanics already in the country in 1980 had grown to 260,232, an increase of just under 100% in a decade. In other words, a significant group of Latino households are moving up the socioeconomic ladder. I must note, however, that the numbers and growth of Mexican middle-class households are much lower and slower than for Latinos as a group (Table 6.11). Also, middle-class Hispanic households

TABLE 6.11. Numbers of middle-class households in 1980 and 1990 in California

Household Type	1980	1990
All households	8,644,600	10,385,000
All native-born households	7,274,200	8,064,247
Native-born Hispanic households	623,040	743,228
Foreign-born Hispanic households	568,900	1,044,423
Middle-class households		
All native-born	3,434,140	5,250,914
Hispanic native-born	211,020	342,757
All Hispanic foreign-born	132,340	260,232
Mexican native-born	157,620	282,165*
Mexican foreign-born	91,280	149,541*

*Arrived before 1980.
Source: U.S. Bureau of the Census, 1983, 1992.

were just 13% of all middle-class households in 1990, barely up from 10% in 1980.

To put these findings in context, the number of all Hispanic households increased by 50% between 1980 and 1990. In other words, total Latino members of the California middle class registered a smaller gain than is reflected in calculations of gains in middle-class membership for a specific group of the Latino foreign-born. Thus Latino households made only modest relative progress in comparison with the California population as a whole. The Rodriguez (1996) results must be treated with caution, even though some Hispanics clearly are undergoing socioeconomic assimilation into the middle class.

The middle-class Hispanic communities are suburban communities. In Los Angeles, Hispanics have established enclaves in the San Fernando Valley, in Montebello, and in Hacienda Heights. Average property values in Montebello and Hacienda Heights are almost $250,000; half of all Hispanic homes in these suburbs are valued at over $175,000. These substantial property values are supported by average incomes of more than $36,000 in Montebello and almost $52,000 in Hacienda Heights. These communities do not have large numbers of very wealthy households; indeed, the middle-class residents are very similar to one another in their incomes. They also behave like similar middle-income white households, supporting political candidates and lobbying for neighborhood improvement and protection.

MOVING AND MELTING

What is the evidence for suburbanization associated with upward economic mobility? Are the new immigrants following the path of other migrants who left the inner city as they moved up the socioeconomic ladder? This position is supported by the evidence that those who moved out of the city were much more likely to be homeowners, but what of other measures of assimilation: language, occupation, and naturalization?

The following analysis compares those who move *within* the inner city with those who move *out* to the surrounding counties. Are those who move out more likely to be more highly skilled in English, to have professional occupations, and to be citizens? The explicit argument is that those who have "made it" or who are "making it" are also making the decision to move to suburban locations, and in doing so are more likely to leave segregated inner-city housing environments and enter less-segregated suburban neighborhoods. Insofar as they also become citizens, they are entering the American mainstream.

In keeping with the arguments about the possible differences between ethnic and racial patterns of assimilation, I examine Hispanics and Asians separately. Also, because of the very large Mexican presence in the Hispanic ethnic group in California, I separate Hispanics into those with Mexican ancestry and other Hispanics. To reiterate, the issue is whether those Mexican, other Hispanic, or Asian movers who move out to suburban locations are initiating the process of assimilation and integration. I do not divide the movers into recent versus older immigrants or second- and third-generation versus first-generation immigrants; that differentiation is not the test in the present case. Clearly, those who were born in the United States or who have lived here longer earn higher incomes and possess better skills, and are more likely to integrate.[1]

We find distinct differences between the skill levels of those who move within Los Angeles City and those who move to the surrounding suburban counties (Figure 6.3). All suburban locations receive a higher proportion of Mexican movers who speak English well, and most receive a higher proportion with professional occupations. The pattern is less clear for occupations; in the case of moves from Los Angeles City, those who move to Riverside do not consistently have professional occupations. These results are attributable either to the difficulty of correctly classifying professional occupations or to Hispanics' general lack of penetration of the professions. The differences for non-Mexican Hispanics movers are even clearer.

The patterns are replicated for Asians in both Los Angeles and the Bay Area. To that extent, they provide a counter to Glazer's (1993) argument that nonwhites do not assimilate successfully. The structural relationships for migrants of Asian heritage are clear, and support the arguments favoring upward and outward integration confirmed for Hispanic and Mexican movers. Without exception, the proportion of higher skill migrants and those with professional occupations is significantly larger in the suburban counties for intercounty movers than for movers within San Francisco (Figure 6.3). The structure of Asian relocation particularly supports the thesis of upward and outward economic integration.

Overall, the skill levels and occupational success rates of Asian-heritage movers are significantly higher than those of Hispanic- and Mexican-heritage movers. Even so, the distinction between within-city movers and those who move to the suburbs is quite clear. Fairly consistently, the inter-regional moves aggregate to a pattern of higher skill level and higher professional status rates than do moves within the region. How is this pattern played out in levels of citizenship as a measure of integration?

Using the measure of citizenship as a measure of societal integration, we find that only about 40% of the intracity movers are citizens, but that the corresponding proportions of intercounty movers range from 48 to 61% (Figure 6.4). These measures of citizenship also must be seen in the context of an ethnic group (Mexicans and Hispanics generally) whose members have been slow to become citizens. The counties surrounding Los

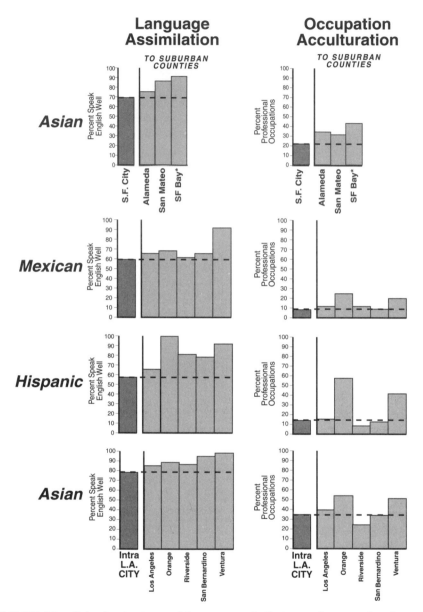

FIGURE 6.3. Suburbanization and language assimilation and occupational acculturation. *Source:* U.S. Bureau of the Census, 1992.

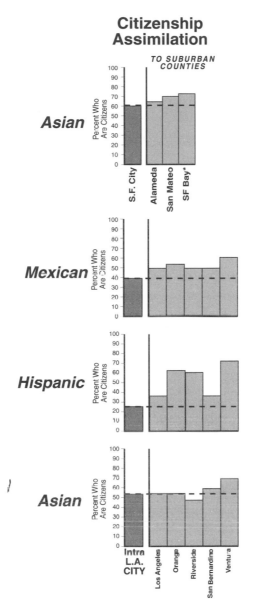

FIGURE 6.4.   Suburbanization and citizenship. *Source:* U.S. Bureau of the Census, 1992.

Angeles and San Francisco, however, are notable for their much higher levels of citizenship. The interpretation of this data seems to support the optimistic notion of a melting pot involving upward socioeconomic mobility and residential suburban relocation.

The data on levels of integration further bolster the notion of assimilation. Residential separation is higher in the core cities of Los Angeles County than in the surrounding, more suburban counties and has decreased in most suburban counties, especially for Hispanic/black residential separation.

The dissimilarity index is often used to measure changes in the degree of segregation in an urban area. This measure shows the degree to which two groups live separately from one another. If the index tends toward 100, the separation between any two ethnic or racial groups is complete. In effect the two groups live in mutually exclusive neighborhoods. As the index tends to 0, the groups are more fully integrated (as in the extreme case of men and women). The index is usually measured at the census tract level because these units are often used as representations of neighborhoods. The details of separation or integration vary by place and time, but at the census tract level the levels of black/white segregation are lower today than in the past; they are about the same for Asian/white separation; and they have increased slightly for Hispanic/white separation (Figure 6.5).

Despite the positive evidence of lower levels of separation in the suburbs, the evidence of increasing segregation in inner-city areas raises questions about the future of integration. In addition, other studies have shown that after an initial decline in the levels of separation, the very large numbers of Hispanics may be stimulating a resegregation (Rolph, 1992).

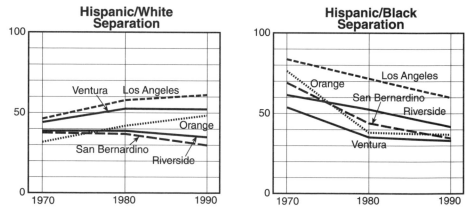

FIGURE 6.5. Changes in separation levels over time in Southern California. *Source:* Clark, 1996a.

Two scenarios are possible, as I have argued elsewhere (Clark, 1996a). From the optimistic's point of view, the way in which the residential mosaic is evolving will lead to greater integration and more mixing. Alternatively, the new mixing, as suggested by the indices, merely reflects a transition from predominantly white to predominantly Hispanic and Asian neighborhoods. In yet a third scenario, integration will increase in the suburbs but it will decrease in the inner-city areas, where most new migrants are settling.

For new immigrant groups, separation from the non-Hispanic white majority is often extreme. The very highest levels of separation occur between Middle Eastern groups and certain Southeast Asian groups. Central Americans also are quite separated from Middle Eastern groups: for example, the dissimilarity index for Armenians and Cambodians is 96 (Allen and Turner, 1997). These very high levels of separation reflect strong cultural and economic differences and the presence of very recent arrivals. The figures illustrate the importance of ethnic havens in creating new residential environments. In addition, the individual ethnic groups' levels of separation seem to be higher in Southern California. By contrast, we find very high levels of mixing between Asian Indians and other groups and between Filipinos and other groups.

The most diverse areas in both Northern and Southern California are in the "rings" around the central cities in Los Angeles and in strips north and south of the Bay Area. These highly diverse areas, where the major ethnic and racial groups are mixed almost equally, fall between the relatively homogeneous inner-city core of blacks and Hispanics and the white coastal regions and northern valleys (Clark, 1996a). The new communities in the center of the city show very low diversity (they are almost totally Hispanic), as do the expensive coastal communities (they are almost totally non-Hispanic white). The neighborhoods of East Los Angeles and Maywood (Hispanic), on the one hand, and Laguna Beach and Newport (non-Hispanic white), on the other, are the least diverse communities in Southern California. The most diverse areas in Southern California include communities in Long Beach, West Covina, Rowland Heights, and Gardena.

For some ethnic groups, separation is increasing at a time when white/black separation is slowly decreasing. This fact reflects both the rapid demographic changes in the state's urban areas and the continuing tendency to congregate with one's own kind, which has always underpinned the urban mosaic. Does this mean that assimilation is slowing? Does residential separation signal less integration?

The results suggest a mosaic in flux. As I noted earlier, the high levels of in-migration have created a situation in which ethnic groups cluster in groups to protect and enhance their status. This is no different from the establishment of enclaves by ethnic minorities at the turn of the century; in themselves the relatively high levels of separation are to be expected. The question is whether the new enclaves are communities through which newcomers will pass gradually on their way to the mainstream, or are they permanent ethnic settlements in which residents live out their lives apart from other groups. Will they be home to a procession of people coming and going, or will they be neighborhoods permanently separated along ethnic lines and embedded in the geography of the large cities? The answer will depend on a combination of changes in the migrants' socioeconomic status and the levels of their own-race/ethnic preferences. If little economic advancement occurs, own-race preferences will ensure concentration.

DOES CALIFORNIA EXEMPLIFY A MULTIETHNIC SOCIETY?

The findings of this extensive discussion of language acculturation, intermarriage and changing socioeconomic status reiterate those from other, earlier studies of ethnic upward and outward mobility. They offer additional evidence that out-migration from core communities, even measured at the general county level, is associated with increased skill levels, increased socioeconomic status, and, most important, maintaining geographic integration of culturally distinct groups. Although integration, in the long run, may be affected by the size of the flows, both Hispanics and Asians in suburban Southern California counties are more fully integrated than their counterparts in the core regions. An optimistic interpretation of the analysis made in this chapter is that the melting pot, although it is promoting blending rather than melting, remains an applicable metaphor for understanding late-20th-century America. In this chapter I offered some evidence of continuing integration, which others have noted for western metropolitan regions (Farley and Frey, 1994).

However, in a less optimistic interpretation, California is regarded as simply in transition to what Allen and Turner (1997, p. 252) call a "collection of ethnic societies." The earlier waves of immigrants, predominantly white, were much more similar to one another than to blacks or Asians; that issue underlies much of the turmoil over recent migration. By the second generation, with little difference in skin color and few strong visual clues to impede intermarriage, a great deal of mixing had occurred. That mixing with the white majority has never occurred for blacks and is doing so only slowly for Asians. This point suggests that the next stage of assimilation will take much longer than the first stage. In addition, demography will play an important role. There is evidence that mixing and assimilation slows in the presence of very large numbers of one or more ethnic groups.

If the flow of Mexican immigrants continues at the current pace, assimilation will be much slower than in previous decades. Mexican immigrants' ability to survive without English, to easily find a spouse of a similar ethnic background, and even to work for a firm owned by fellow ethnics will slow the process of change. The issue is not whether this is good or bad; it is simply a fact of the process of change in the context of continuing large-scale flows. This situation suggests, however, that the assimilation of the future will be different from that of the past, and it is possible that the process could usher in an era of separation and balkanization.

Although the evidence for separation and division is not clear, some individuals and families appear to be leaving areas of high immigration.[2] Yet because these areas are also the central cities of large metropolitan areas, the out-migration is more than simply a response to new and increased levels of immigration. It is also a reaction to restructuring and employment changes, to crime and perceptions of deteriorating public safety, and to the more nebulous tensions related to urban living. Even so, if the differential patterns of immigration and out-migration continue, it is entirely possible to envision a completely changed and ethnically separated metropolitan structure in which new immigrants are crowded in the center, while postindustrial (white) employees are located in suburbs and small towns. Such a balkanized picture of the California contrasts sharply with the cultural diversity and vibrant ethnic mixing that have been so essential a feature of our nation's past.

The tentative evidence shows the likelihood of very strong differences between levels of separation in the metropolitan areas of the high immigrant states, such as California, New York, and Florida, and the cities of the low immigrant states in the Northeast and Midwest. The emergence of such divisions will create a cleavage between new, young ethnic populations and older and white native-born populations, and will invite scenarios of conflict and contention when the new young groups are asked to fund the Social Security of aging white populations. Even within the ethnically diverse states, the division between the new working-class poor and the wealthy native-born whites is a further outcome of the recent high levels of immigration. Immigration levels and preferences will have major effects on the future of division and separation. Although the processes of change are underway, the outcomes are less clear. I return to these issues in my final chapter.

NOTES

1. Obviously, the issue of cohort differences may affect the rates and success patterns of out-migration. The sample data are not rich enough to allow a detailed

breakdown by all immigrant groups and by year of entry. Even so, the average patterns of upward and outward mobility would be enriched only by noting which groups have the greatest probability of making the transition, and would not alter the general argument advanced with this analysis. For additional complexity, however, the analysis can be segmented for foreign-born and native-born intercounty movers. The pattern whereby intracounty movers have lower socioeconomic status than intercounty movers holds for both foreign-born and native-born movers. However, two points are notable. First, as expected, the native-born movers are significantly higher up the socioeconomic scale than the foreign-born movers. Second, the difference between intra- and intercounty movers for the foreign-born is somewhat greater than for the native-born migrants. In other words, those foreign-born movers who make the upward and outward move are more successful than those who do not or can not make the move.

2. The metropolitan magnets of Los Angeles/Riverside, San Francisco/Oakland, and San Diego all had large immigrant inflows. The domestic outflows in the period 1990–1995 were more than a million for Los Angeles/Riverside, and a quarter million for San Francisco/Oakland. The loss from San Diego was 140,000 (Frey, 1996b).

Chapter 7

THE POLITICS OF IMMIGRATION

As ethnic diversity has increased in California and especially in its metropolitan areas, this diversity has refocused attention on immigration in general and on the impacts of immigration on communities and neighborhoods in particular. The state government is concerned with the costs of providing services to immigrants, both documented and undocumented. The public feels ambiguous about how their society is changing. Immigrant support groups wish to keep the current patterns of support available for the new immigrants, but these groups have varying political agendas that often clash with one another (Skerry, 1993). Even the public, despite its concern about large-scale flows, is divided between those who support the increasing diversity of our society and those who want to keep the status quo.

Political reaction to the immigration problem is as varied as the immigrant groups themselves. But it seems to be generally agreed that undocumented immigration should be controlled (U.S. Commission on Immigration Reform, 1997). In this chapter I will examine the range of political reactions to large-scale immigration, from the perspective of both immigrants and the native-born. I believe that these reactions are far more complex than descriptions of immigrant bashing and nativist xenophobia would suggest.

THE PUBLIC VIEW

Nationally, an increasing proportion of the population favors a decrease in immigration. In 1965, when the Hart-Cellar Act, a turning point for immigration flows, was enacted, only 33% of the American public thought that immigration should be decreased. Most people did not perceive any problem then with the level of immigration, and consequently saw no need to change U.S. policy. By 1993, however, 65% of the U.S. population wanted a decrease in immigration levels (Gallup Poll Organization, 1993). Today, opinion is divided almost evenly as to whether immigrants are a positive (44%) or a negative force (49%) in our society (*Wall Street Journal,* June 27, 1997 p. B1). However, the averages mask strong differences by income. Among blue-collar workers earning under $20,000, for example, only one-third thought that immigrants had a positive effect. In contrast, the well-off, those earning more than $75,000, were much more positive about immigrants' contributions.

Responses in California mirror the responses nationwide. In general, California's population thinks that immigration levels should be reduced. Again, those with lower incomes and less education are much more likely to hold this view (Table 7.1). In a 1993 poll, almost two-thirds of respondents with a high school education wanted to curtail immigration flows; this was also true for 58% of the black households and 53% of the white households. These findings reflect clear concerns of the native-born population, especially the low-skilled native-born population, about the very large-scale flows I examined in the earlier chapters. The findings for questions about refugees yield more consistent responses across groups and generally indicate greater interest in limiting the flows of refugees (Table 7.2). Hispanics as a group had the largest proportion who wanted to expand immigration. Asian households wanted to keep immigration the same and did not want to increase it.

Much of the public concern in California pertains to illegal immigration. For 10 years there has been a consensus that undocumented immigration is a serious problem (Figure 7.1). More than 50% of the population surveyed in California reported that they were "extremely concerned" about illegal immigration (Figure 7.1). If those who said that they were "somewhat concerned" are included, the proportion rises to more than 80%. This opinion hardly differs across ethnic groups, although the strongest concerns about the seriousness of illegal immigration are voiced by those with a high school education and members of middle-income households (Figure 7.2). That almost 80% of most groups regard illegal immigration as a serious problem helps to explain why the public supports

TABLE 7.1. How many immigrants should be admitted legally?

	All	Hisp.	Black	Asian	White	HS Educ.	Middle income	Age >55
Increase	8.1	15.2	5.1	3.5	5.1	5.0	6.9	6.2
Same	32.7	38.5	35.6	47.9	36.9	28.5	36.7	30.9
Reduce	49.3	40.3	57.6	31.3	53.0	64.0	52.0	57.5
No opinion	4.9	6.1	1.7	4.2	5.0	2.5	4.4	5.4

Source: Field Institute, 1993.

TABLE 7.2. How many refugees should be admitted?

	All	Hisp.	Black	Asian	White	HS Educ.	Middle income	Age >55
Increase	8.3	10.8	8.5	10.4	7.5	5.0	7.3	5.4
Same	27.8	35.1	23.75	33.3	25.8	23.0	26.4	22.0
Reduce	57.9	48.1	62.7	50.0	60.8	66.9	61.5	66.0
No opinion	5.9	6.1	5.1	6.3	22.6	5.0	4.9	6.6

Source: Field Institute, 1993.

harsher penalties for illegal immigrants but is generally favorable to the immigrant population as a whole.

In general, illegal immigrants are viewed as exerting an unfavorable effect on California (Figure 7.3). Curiously, however, most respondents do not think that illegal immigrants take jobs away from other California workers (Table 7.3). Again, these responses vary greatly by ethnic status, education, and income. Black households feel strongly that illegal immigrants take jobs away, as do households headed by those with a high school education. Not surprisingly, Hispanic households overwhelmingly

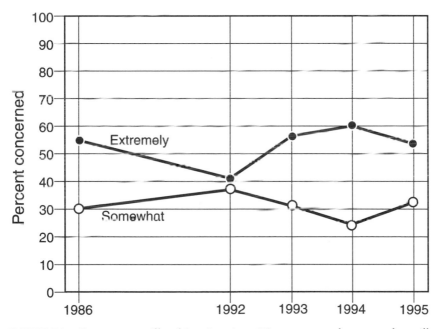

FIGURE 7.1.   Responses to illegal immigration. (How concerned are you about illegal immigration?) Source: Field Institute, 1993.

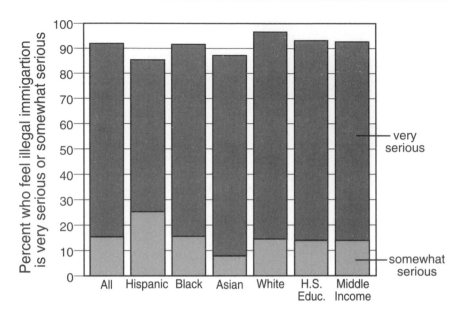

FIGURE 7.2.   Public opinion on illegal immigration in California. (How serious do you think illegal immigration is in California?) *Source:* Field Institute, 1993.

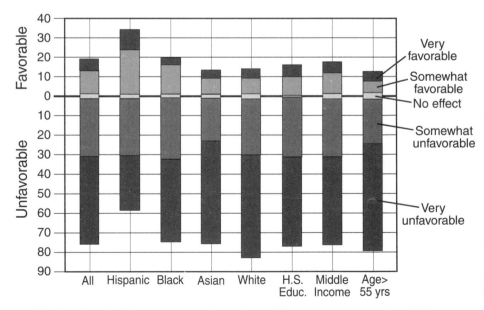

FIGURE 7.3.   Public opinion on the effect of illegal immigration in California. *Source:* Field Institute, 1993.

TABLE 7.3. Do you think that illegal immigrants take jobs from other Californians?

	All	Hisp.	Black	Asian	White	HS Educ.	Middle income	Age >55
Yes	35.9	26.0	54.2	33.3	42.6	51.5	40.9	44.4
No	52.3	68.8	33.9	60.4	47.4	40.6	51.3	43.6
Don't know	8.8	5.2	11.9	6.3	9.9	7.9	4.1	12.0

Source: Field Institute, 1993.

think the opposite; Asian households' views are similar to those of Hispanics. White households are divided on the issue.

Although public opinion about immigration in general is divided, it is less deeply divided about illegal immigration. While public responses show a strong desire to reduce immigration, they also indicate strong support for new immigrants' customs and language. Seventy-two percent of those polled think that immigrants should keep their customs; 81% favor maintaining the immigrant language. The public even expresses strong support for providing emergency health care for illegal immigrants, though a bare majority do not think that illegal immigrants' children should be enrolled in the school system (Field Institute, 1993). These divisions reflect the clash between the liberal and humanitarian values of the American population, on the one hand, and their worries about society changing for the worse, on the other hand. To cast these complex responses as racist or anti-immigrant is to miss the nuances of the debate about how many immigrants our society can absorb without destabilizing its social structures.

Surveys of Latinos themselves contradict the stereotypic belief that these immigrants' views are different from those of the population as a whole. Although Hispanics in particular want to keep the doors open to immigrants (a natural response), all immigrants express views similar those of Californians as a group about education, crime, and the economy. Latino immigrants list education and the economy among their major concerns (Rose Institute, 1988). In a survey designed to address Latino concerns, 58% of the respondents favored making English the official language, 86% believed that welfare recipients should be required to work, and 80% believed in the American dream of being able to get ahead regardless of ethnic background or national origin (Rose Institute, 1988). Significantly, 53% believed that immigration laws should be strictly enforced, including laws concerning the deportation of undocumented aliens. The survey respondents were nearly 40% foreign-born.

As I stated earlier, polls have shown widespread concern about levels of immigration, and especially levels of illegal immigration, for more than a decade. Much of that concern found expression in the debate over Proposition 187, a referendum hailed as "Save Our State" by its proponents. Designed to remove all funding for illegal immigrants, Proposition 187 would have required teachers and health workers to report undocumented aliens to the Immigration and Naturalization Service. Proposition 187 drew both fervent approval and heated protest. It was used as a cornerstone of the 1994 reelection campaign of California governor Pete Wilson. Though it passed by almost 60% of the vote, it was later declared unconstitutional by a federal district court judge. The case will next be argued before the federal court of appeals.

PROPOSITION 187 AS A WINDOW ON REACTION TO IMMIGRANTS

The vote on Proposition 187 is a window on the public reaction to large-scale immigration, and offers some insights into anti-immigrant sentiment and "nativism." Analysis of the vote shows that the reaction to immigration is more complex than suggested by the media and liberal politicians who decry racism and immigrant phobia. The voting patterns concerning Proposition 187 confirm the opinion polls' findings of widespread support across ethnic groups for limiting immigration. The vote also hints at the complex alliances that have formed in regard to continuing the high levels of immigration. The strongest pro and con views about the size of the immigrant flows also coalesced around Proposition 187.

On the one hand, Proposition 187 has been viewed as an attempt by conservatives to assert state control over the nation's borders, and even for forcing the federal government to recognize the local outcomes of continued large-scale migration into California. On the other hand, Proposition 187 is seen as an attempt to "criminalize people of color, suppress their culture, and divide them along class and racial lines" (Cervantes, Khokha, and Murray, 1995; Garcia, 1995).

Proposition 187 was designed to remove funding from "illegal" (i.e., undocumented) immigrants, and specifically to exclude undocumented children from schools and colleges, to deny nonemergency health care to undocumented residents, and to require the police to verify the immigrant status of all persons arrested. Proposition 187 did not affect legally admitted migrants. Proposition 187 passed everywhere but in seven Northern California counties in the San Francisco Bay area, a traditionally Democratic and liberal set of counties centered on the cities of San Francisco, Berkeley, and Oakland (Figure 7.4). The November 8, 1994, vote was followed 2 days later by eight lawsuits that petitioned the courts to enjoin enforcement of the proposition. The provisions of the proposition were immedi-

FIGURE 7.4.   The vote for Proposition 187 by county in California. *Source:* Clark, 1998b. Reprinted by permission of John Wiley and Sons.

ately blocked, and soon thereafter the federal court denied its constitutional validity.[1]

The vote on Proposition 187 attracted interest outside California, indeed, outside the United States. Government officials from Mexico and El Salvador condemned the vote as intolerant and an attack on human rights (*Los Angeles Times,* 10 November 1994, p. A28). Of course, neither Mexico nor El Salvador is noted for exemplary records for upholding human rights. In fact, these countries' response to Proposition 187 was less a reflection of concern for human rights than a sign of fear: fear of the consequences of the sudden return of populations for whom there are neither jobs nor public services. Indeed, Mexican government officials have frequently expressed concern about the potentially devastating effect on Mexico's impoverished countryside if several hundred thousand immigrants were to return from the United States.

The popular opposition to Proposition 187 was fueled in part by questions about its constitutionality (Margolis, 1995). Earlier federal court rulings emphasized that immigration is a federal matter, not the province of state laws or state control. Thus it is unconstitutional for states to make laws regarding immigration. A second challenge to Proposition 187 relates to the federal law requiring that states provide a free public education to all children. Among the arguments used in the lawsuits against the proposition, the plaintiffs cited *Plyler v. Doe* (1982), a Supreme Court ruling that prevented the State of Texas from blocking the use of education funds for illegal immigrants. A third challenge to the proposition comes from the Fourteenth Amendment to the Constitution, which expressly guarantees individual freedoms for all persons in the United States. The Supreme Court has ruled that all persons in the United States are entitled to due process and equal protection, even if they are in this country illegally. Margolis (1995) and Pasqualucci (1994–1995) explain the legal background to the challenges.

INTERPRETING THE VOTE

Proposition 187 passed by a vote of 59% "for" to 31% "against." According to exit polls, non-Hispanic whites voted 63% in favor. African Americans and Asians voted 56% and 57% in favor (*Los Angeles Times,* 9 November 1994, A1). Most Hispanics, however, opposed the proposition; only 31% voted in favor. These aggregate numbers mask considerable local variation in the vote. I have evaluated that variation in one county by using data from the Election Results Rental File for 1994 of the County of Los Angeles. The 6,104 voting precincts have been aggregated to 1604 census tracts of about 4,000 to 8,000 persons each, which approximate neighborhoods in Los Angeles County. Since we also know the ethnic composition of the tracts and the percentages of the population registered as Republicans and Democrats, we can provide an interpretation of how different communities voted on the proposition.

The analysis of the vote by city and community shows very different patterns of support and opposition within Los Angeles County (Figure 7.5). The box-and-whisker diagrams measure the range of support over the tracts within a city or community. The median tract vote was about 68% in favor of the proposition in Burbank, but only about 22% in favor in East Los Angeles. This finding is hardly surprising, although many ethnic communities and some Hispanic areas actually supported the proposition. The 15 cities and communities depicted in Figure 7.5 show the variation in responses from suburban cities (Burbank and Glendale in the north and west, Whittier in the east), inner-city black communities (South Central, Compton), inner-city Hispanic communities (Bell, East Los Angeles), and wealthy cities (Santa Monica, Beverly Hills). In the valley communities

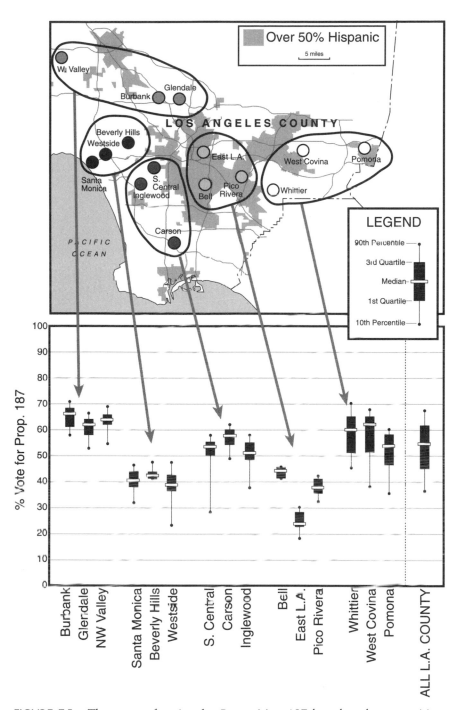

FIGURE 7.5. The range of voting for Proposition 187 by selected communities. *Source:* Clark, 1998b. Reprinted by permission of John Wiley and Sons.

(Burbank, Glendale, West Valley), the voters in the median tract voted strongly in support of Proposition 187. The 90th percentile tract (the upper dot) voted about 70% in support; the lowest, the 10th percentile tract, voted approximately 57% in support. Three-quarters of all the tracts voted more than 60% in support of the proposition.

Apart from East Los Angeles, inner-city Hispanic communities showed surprisingly high levels of support for the proposition. In Bell, the median tract was about 43% in support; the whole city ranged between 40 and 47% in support. The pattern of support varies from opposition, in higher status and more affluent communities on the West Side of Los Angeles, to strong support in the Valley communities. Why was the vote against Proposition 187 so strong on the West Side of Los Angeles? Was this a civil rights response to perceived repression or a self-interested vote for keeping low-paid domestic workers?

It has been suggested that as ethnic groups assimilate, thanks to longer residence in the United States and movement to more suburban locations, their support for controlling immigration will increase (Skerry, 1993). This idea is supported by some data relating to the vote for Proposition 187. For example, suburban Latinos, such as those who had moved to the San Fernando Valley, were more conservative (i.e., pro-Proposition 187) than Latinos from inner-city areas (*Los Angeles Times*, 12 May 1996, B2). Tracts containing larger proportions of Hispanic registered voters showed greater support for the proposition (Clark, 1998). The results are also consistent with immigrant-citizen voters who are concerned about the effects of large numbers of undocumented immigrants working for very low wages. Even in majority Hispanic neighborhoods, 43% of the voters supported the proposition; indeed, more than one-quarter of all majority-Hispanic tracts voted in favor.

The average vote in majority-Asian neighborhoods was 52% in favor; three-quarters of all these neighborhoods voted in favor of the proposition. These figures are quite consistent with earlier survey results, which report a continuing concern about illegal immigration among all immigrant groups. The results suggest the need for caution in identifying the vote on Proposition 187 as mere immigrant bashing.

Patterns of voting correlate positively with variables that measure the percentage white in the neighborhood, and with median income and percentage Republican. These results suggest a simple anti-immigrant position in a large portion of the population; the simple correlations are consistent with our expectations about the behavior of groups.

An integrated model of the variation in support for Proposition 187 suggests a more nuanced interpretation of the vote (Table 7.4). All the variables are significant and, as expected, the strongest explanatory variable is the proportion of Republican voters. The percentage black, however, is

Table 7.4. Regression coefficients for percent vote yes
on Proposition 187 and socioeconomic variables

Variable	Coefficient
Percent white	.06
Percent black	.32
Percent Hispanic	−.22
Median income	.07
Percent of pop. <30 years	.09
Percent of pop. >54 years	.11
Percent with high school	.30
Percent Republican	.79
Adjusted R2	.72

All variables are significant at the 0.5 level. Interaction effects were not significant and are not reported.

also a strong predictor, as is the proportion with no more than a high
school education. Thus the greater the proportions of either Republican
voters or black voters, with income and age held constant, the higher the
vote for Proposition 187. These results are consistent with nativist sentiment by conservative Republican voters, and the votes of a "threatened minority," in the case of tracts with high proportions of black voters or persons with less than a high school education. The votes by Asians may
reflect both reactions to competition and the tensions that are reported
among all recent ethnic immigrants (Abelmann and Lie, 1995).

The reaction to immigration as deduced from this analysis of actual
voting behavior suggests a complex local response, one that cannot be cat-
egorized simply as nativist and racist reaction. The strength of cultural diversity depends on the delicate balance of competing groups. The reaction
to Proposition 187 can be viewed as a response to a perceived disturbance
in this delicate balance and a consequent fear of divisions fueled by ethnic
rivalry. The data on voting behavior also suggest that local Californian
concerns about mass migration are at odds with national policies governed
by diverse agendas, including civil libertarians' concerns about a national
commitment to the role of the United States as a nation of immigrants, and
big business's (especially agricultural interests) desire for open immigration
as a continuing source for low-cost labor.

ELECTORAL REPRESENTATION AND POLITICAL POWER

In 1965 the California Assembly contained one Hispanic representative, no
Asians, and a handful of blacks. Clearly, minority groups were underrepre-

sented. The Federal Voting Rights Act of 1982 properly redressed inequities in African-American voters' access to political representation and power. The creation of single-member rather than multimember districts, as well as the prohibition of "packing" (concentrating minority voters into one district) and "cracking" (dispersing black voters over several districts so that they would have little influence), helped create opportunities, first for blacks, and later for Hispanics and Asians, to win local, state, and congressional elections.

Today significant numbers of black representatives are found at all levels of government. The numbers of Hispanics and Asians are increasing, although these groups are still underrepresented. The Hispanic caucus in the California legislature has two dozen members, and several Asian Pacific representatives hold office in the state government. California has 682 Hispanic elected officials in various position in the federal, state, and local governments (NALEO, 1992), including members of the U.S. Congress, state senators and representatives, and an array of local politicians. More than 200 Asians are serving as elected officials, but far fewer in state and federal elected offices (National Asian Pacific American Political Almanac, 1996).

The continuing drive to redistrict in order to emphasize ethnic representative raises an important question: To what extent should ethnic identity be the basis of reorganizing political power (Clark and Morrison, 1995)? What happens when political districts are restructured to accommodate new groups (notably Asians and Hispanics) that may include very large numbers of recent (and sometimes undocumented) new residents? Of course, ethnic groups in the past pressured the government for representation. According to one view, the current drive for political representation based on ethnicity is simply a variant of earlier pressure by Irish, Italian, and Polish immigrants in large eastern cities in the late 19th century. But other observers ask whether the struggle for power *between* ethnic groups, especially blacks and Hispanics, is producing more political influence for minorities or instead just more racial tension, conflict, and eventual balkanization (Clark and Morrison, 1995; Schlesinger, 1992).

In a new context involving many competing groups, there are compelling arguments for rethinking how we create voting districts. I, for one, think the time has come to create districts based on shared powers by multiethnic groups rather than on the dominance of one particular group. Latinos take varied routes to local political empowerment, depending on their numbers and characteristics (Morrison, 1998). The issue can be approached in a number of ways, but the fundamental questions are whether we believe that *only* a minority representative of a particular group can speak for that group and how immigrant groups should be recognized in political redistricting.

In almost all cases of redistricting, areas with equal populations have been used to create U.S. congressional, state legislative, and local electoral districts. In most states, the use of total population or some other measure (such as the number of citizens or the number of voting-age citizens) as the basis for apportioning electoral districts would have little impact on the structure of the districts. Outcomes in high immigrant states, however, may well reflect populations in transition, especially in situations involving large numbers of undocumented immigrants and noncitizens in general. Then the issues of who is a citizen and what proportion of the total population are citizens become important parts of the political process.

It is possible to illustrate the varying effects of citizenship, voter registration, and voter participation by examining Hispanics' involvement in the election process, controlling for age and documentation. Whether the cause is large numbers of undocumented immigrants, legal immigrants' inexperience with the democratic process, or lack of interest, the outcomes are dramatic in terms of participation in the political process. We cannot learn the precise number of Hispanics who vote in a particular election (because voters do not identify themselves by ethnicity when they vote), but by matching the Spanish-surname census file with the voter-surname file we can estimate the number of Hispanics who voted in recent elections in Los Angeles County.

The results of this matching show that Hispanics have a relatively low turnout, reflecting one of the lowest of all groups' commitments to the electoral process. In four recent elections, the total registered voter population in Los Angeles County was about 3.5 million. The Hispanic registered voter population was between 350,000 and 430,000. This number is increasing, but the Hispanic proportion of all registered voters has remained relatively constant at about 12%.[2] Thus Hispanics constitute 37% of all Los Angeles County residents and 33.3% of people 18 and older, but only 19.8% of voting-age citizens. To distinguish further, 17.2% of Hispanics are voting-age citizens who speak English well, and only 14.1% were registered to vote in 1990 (Figure 7.6). This low rate of participation can be explained at least in part by the small number of citizens relative to the voting-age population, but it also reflects the proportion of voting-age Hispanic citizens who are comfortable with English (Figure 7.6).

In the long run, immigration may make politics the area of greatest social tension in our society. Compton is a classic case study of the impact of ethnic change, specifically ethnic change fueled by international migration, on a locality. In 1970 Compton had become a majority black city, with blacks making up 70% of the population. By 1990, when the census counted slightly over 93,000 persons, Compton was still majority black, but barely so. Now the city is 55% black, 42% Hispanic, about 2% Asian/Pacific Islander, and 1% non-Hispanic white. In 1994 the Compton

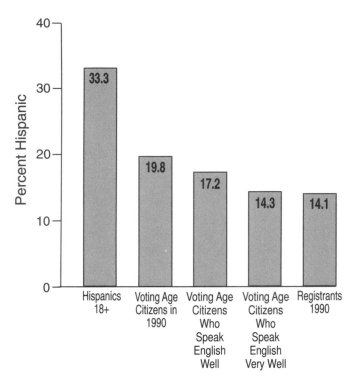

FIGURE 7.6. Hispanic political participation, ages 18+ in Los Angeles County. *Source:* Clark and Morrison, 1995. Reprinted by permission of the Population Association of America.

school district was 59% Latino, 39% black, and 2% Asian, Pacific Islander, and other. The higher proportion of Hispanics in the school system is the effect of the age shift whereby younger immigrant families enter inner-city communities.

In the 1960s and 1970s, blacks in Compton struggled first to win political power and then to win control of an increasingly demographically black city. Now Latinos in Compton are alleging discrimination by blacks (*Los Angeles Times,* 21 August 1994, p. A1). This is a classic case of quota thinking, one that offers a persuasive argument favoring the "influence approach" rather than the "dominance approach" to political and social organization. Hispanics point out that they are underrepresented in government (just as blacks did 30 years ago). Compton has about 530 employees, who are 78% black, 11% Hispanic, 8% white, and slightly over 3% Asian.

The slowness in the power shift from black to Hispanic can be explained by the citizenship and naturalization process. Although Compton's population is close to 100,000, there are only 34,243 registered voters. Fewer than 18% of this number voted in the 1994 primary election, and few of these were Hispanic voters. This distribution reflects the fact that Compton, like Los Angeles County as a whole, contains very large numbers of undocumented immigrants in addition to Hispanics who are present through immigration amnesty (IRCA applicants); the latter still are not citizens and cannot vote. But even if this situation explains the participation outcomes, it does not deal with the increasing tension between black and Latino groups. Black leaders view the Latino's push for power as one that utilizes the gains of the African-American civil rights movement to empower a population that is white. The issue, however, is larger than a dispute in Compton; it represents one possible future for communities in Southern California, and eventually in the nation. It is an issue of political upheaval as groups struggle for group power rather than for individual influence.

VOTING DISTRICTS

The debate about the structure of districts became more critical after passage of the amendments to the Voting Rights Act in 1982 and the *Thornburg v. Gingles* Supreme Court decision in 1986. The decision in *Thornburg* was an attempt to state explicitly the rules for electing representatives. The Court reiterated the principles of electoral equality. It stressed the principle of one person, one vote, and the idea that the vote by a person in one district should carry the same weight as the vote by a person in another district. The Court also ruled that each legislator should represent the same number of persons as every other legislator; this is the principle of representational equality. Third, the Court ruled that there should be no dilution of the minority vote. That is, if a minority population could elect a representative of their choice, presuming that minority population was sufficiently geographically cohesive, they must be given the opportunity to do so through effective voting.

Obviously, if the proportions of citizens coincide with the proportions of voters, drawing districts by population and by citizens would have the same effect. Rapid in-migration, however, with large numbers of undocumented aliens and noncitizens, can create anomalies.

Again, specific local outcomes and effects can be illustrated by using the case of Los Angeles County. Redistricting in 1990 created a supervisors' district in Los Angeles that was designed to be equal in population to the other four supervisors' districts (Figure 7.7). In fact, however, these court-ordered districts violated the principle of equal population size when

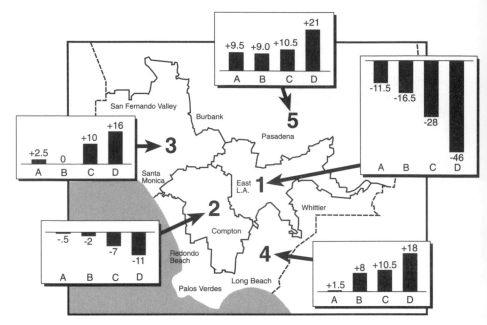

FIGURE 7.7.   District percent variation from equality of (A) population, (B) citizens, (C) voting-age citizens, and (D) registered voters for supervisorial districts in Los Angeles County.

they were drawn (Clark and Morrison, 1991). But even if we assume that the districts were equal in population according to the current standard for redistricting, they differ greatly on the measures of actual political participation.

The court redistricting ostensibly created five districts equal in population, but those districts varied substantially in terms of numbers of citizens, voting-age citizens, and registered voters (Figure 7.7). The variation for citizens occurred largely in the Hispanic district, as expected, where the "shortfall" of citizens was almost 17%. The shortfall increases for voting-age citizens, and is almost 50% for registered voters. These court-ordered districts can be contrasted with a set of districts that are balanced for population, citizens, and voting-age citizens (within some margin, because it is not possible to fulfill all criteria simultaneously). The geographic pattern of districts is quite different from the actual court plan, and it is clear that a majority Hispanic district cannot exist with this formula (Figure 7.8). In fact, this district would split the concentration of Hispanics among several districts, and it is just such a division that the court has ruled unconstitutional. Clearly, it is quite difficult to create districts that both reflect the

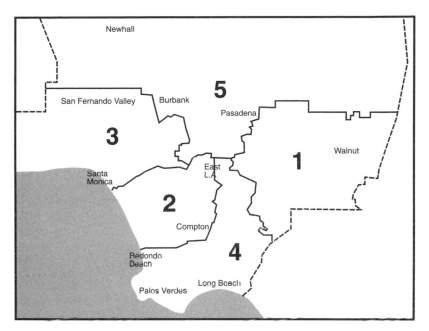

Newhall

San Fernando Valley   Burbank   **5**

Pasadena

3

Walnut

Santa
Monica   Ea t
L.   **1**

2

Compton

Redondo
Beach   **4**

Long Beach

Palos Verdes

FIGURE 7.8.   An example of equal representation districts in Los Angeles County.

numbers of citizen voters and respect the rights of minority groups. One way to accomplish this is not to focus on "ethnic" districts, but rather to attempt to create districts in which each of several groups have a substantial presence without a majority.

In some jurisdictions, demographic realities force a choice between according equal representation to all persons and giving equal voting power to all citizens. The tension is not simply a matter of citizens and noncitizens. An increase in one immigrant group, legal or undocumented, has affects on other immigrant groups and resident ethnic populations. When a relatively disadvantaged immigrant population increases rapidly, as in the case of the Mexican and Central American groups under discussion here, it is likely to do so in inner-city, socioeconomically poorer neighborhoods. If these neighborhoods are home to one or more *other* ethnic or racial group, the stage is set for interethnic conflict. When political power is reorganized, as in the Los Angeles County redistricting, recent in-migration has potential impacts on resident ethnic populations. In such a case, further divisiveness and conflict may result.

At first thought, a head count seems the fairest and most equitable way to allocate representation, and thus power. The idea of using the total number of residents as a basis for appointment is persuasive because it emphasizes equality of access to political representatives. As the result of creating districts based on equal numbers of registered voters, however, some districts would contain much larger populations than others; thus some districts would be precluded from equal access to elected officials. This idea, of course, is based on an idealized view of elected officials as those who serve their total constituency rather than influential organized groups with money and lobbying skills.

In reality, when districts are created based on total number of residents, but contain large numbers of noncitizens who do not vote, power is retained by a much smaller group in some districts than in others; this situation creates advantages for those in districts with fewer registered voters. In the 1991 election for supervisor, for example, the supervisor in the Hispanic district was elected with 55% of 88,102 votes cast. In contrast, in the four other district elections for supervisor, the winning proportions were not much different but the number of votes cast ranged from 170,000 to 270,000. Also, as I have argued elsewhere, there is no guarantee under continuing and intense levels of migration that such processes are either temporary or likely to change soon. The situation may well continue, given the economic and social climate across a proximate border, for example, the border between the United States and Mexico.

A second element of potential conflict relates to the community structures with which I opened this discussion. Boundaries separate or concentrate, isolate or unify, divide or amalgamate a community. Geography matters, and it matters greatly in issues of local control and local power. Thus some districts preserve communities of interest, while others do not. In the debate over the structure of supervisors' districts, some cities felt that the plan "puts us in with cities with which we have nothing in common" (*Los Angeles Times,* 8 August 1990, B4). Creating an ethnic district preserves and reinforces the community of one ethnicity, but at the expense of emphasizing its separateness from other ethnic communities. District 5 of the Los Angeles School Board illustrates the tension involved in creating a district that has been described as one that "snakes like a crooked finger" (*Los Angeles Times,* 27 February 1995, B3) through a variety of neighborhoods (Figure 7.9). Numerous other "communities of interest" are violated by the lack of any intersection between the school district voting area and city and state agencies with which cities and constituents interact.

In the long run, dividing geographic space along ethnic lines to ensure the election or empowerment of different groups may become the most divisive force in U.S. metropolitan areas. The creation of electoral districts for different ethnic groups has been described as political apartheid (*Holder v. Hall,* 1994), a process that reinforces the already strong divisions within our divided society. The problem arises because when two ethnic groups overlap residentially, it is often impossible to fulfill both groups' "majority" desires

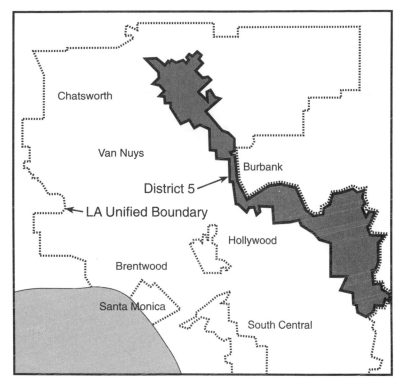

FIGURE 7.9. District 5 of the Los Angeles Unified School District electoral regions.

simultaneously. For example, in two alternative potential community college districts in the Long Beach area in Southern California, it is possible to
maximize either the black population or the Hispanic population (Figure
7.10), but not to do both. An additional alternative is to create districts that
foster opportunities for potential coalitions (Clark and Morrison, 1995).
The creation of "influence" districts rather than "dominance" districts is
one possible solution to the separateness and ethnic division that have been
fostered by the current interpretations of the Voting Rights Act. Moreover,
such influence districts would be increasingly easy to form, as the state and
its metropolitan areas become even more diverse.

OBSERVATIONS: PROTECTED GROUPS, GROUP IDENTIFICATION, AND POLITICAL TENSIONS

Many of the new immigrants enjoy protected status under U.S. civil rights
legislation and the Voting Rights Act. This protection extends from affir-

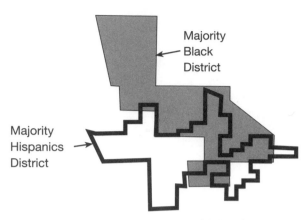

Majority
Black
District

Majority
Hispanics
District

FIGURE 7.10.   Overlap of potential districts to establish either a majority Hispanic
or a majority black voting district in the Long Beach Community College District.
Source: Clark and Morrison, 1995. Reprinted by permission of the Population Association of America.

mative action for university entrance to the creation of an electoral district
when the protected class can be shown to be sufficiently numerous and cohesive to elect a representative of its choosing. But how should the Voting
Rights Act deal with the recent very large-scale immigration? The Voting
Rights Act was designed to deal with the effects of past discrimination and
to redress its insidious effects, particularly on blacks. Few would object to
extending that protection to Asian and Hispanic citizens who suffered discrimination in the 1940s, 1950s, and 1960s. At most, however, we are
speaking of hundreds of thousands rather than millions of new immigrants.[3] It is worth noting that the Voting Rights Act was produced in response to a biracial setting; it was intended to enfranchise blacks in the
South who had been systematically excluded from the electorate. The original purpose of the Voting Rights Act was to eliminate barriers to voting
such as literacy tests and poll taxes; those devices were little more than a
subterfuge to prevent blacks from voting (Davidson, 1984).

     As an additional complication, the identification of group cohesive-
ness is not straightforward. A number of commentators have argued that
ethnicity is a social phenomenon (Bean and Tienda, 1988), and that it can
even be changed by the group members themselves. In some sense, ethnicity is defined by being away from the homeland; thus nationality is defined
as much by that characteristic as by any other. The whole issue of ethnicity
is encapsulated in the debates over defining the Hispanic population. The
very term "Hispanic" groups together new immigrants, second-generation
natives, and refugees; this grouping and definition may owe as much to

Census Bureau classification as to the ethnicity of the group. In the past 30 years, the Census Bureau has used Spanish surnames, language, self-identification, or some combination of these criteria as a way of defining the "Hispanic" group. This classification, however, glosses over the differences within the group; indeed Bean and Tienda (1988) suggest that it is "impossible to speak of a single unified Hispanic population in any strict sense of the term" (p. 398). They point out that Hispanic subpopulations appear to be different from one another and that only a division by national origin and nativity status will capture the group's variability. Conner (1985) also emphasizes the idea that the term "Hispanic" artificially links people of widely different racial, ethnic, and linguistic backgrounds. More than one commentator has suggested that even Mexican Americans are not a single homogeneous group (Conner, 1985; Davis, Haub, and Willete, 1988). This point raises serious questions about the notion of protected groups and their role in the future of our multicultural society. At the extreme, it suggests that the notion of protected status for any group other than black Americans is misguided, and will foster the very same problems among other "protected" groups that it was designed to remedy among blacks.

This position, however, is not accepted by many Latino commentators. Even though they recognize the great diversity within the Hispanic category, they argue that since society treats Hispanics increasingly as a single group, Hispanics are beginning to think of themselves as one ethnic group (Moore and Pachon, 1985). In addition, the number of persons of multiple races has increased notably; they will be measured as such for the first time in the 2000 census. How will these groups be treated in the complex categorization that leads to advantages and disadvantages?

In the case of an increasing divergence between the population as a whole and a growing, underrepresented, but politically active subgroup, the legitimacy of the political structure itself is likely to be challenged, and conflict is certainly a possible outcome. The identification of large numbers of noncitizens who are barred from political participation because such participation is restricted to a different sociological group sets in motion the conflict between the politically empowered and the disenfranchised. Some suggest that this situation calls for rethinking the notion of citizenship, perhaps by moving beyond the legal definition of citizenship (i.e., one based on native birth, period of residency, passing a citizenship test, etc.) to an approach based on community activism and participation (Pincetl, 1994).

In fact, however, the arguments about citizenship are a replay of arguments about quotas and representation that take us back to the initial arguments about space and representation. Thus commentators note correctly that Latinos or Hispanics are underrepresented as voters, office holders, and administrative appointees (Pincetl, 1994), and that these numbers should be adjusted to achieve proportional representation. Enshrining these patterns in quotas and special districts, however, will likely aggravate ethnic conflict rather than create ethnic cooperation.

The survey data and the polls emphasize that a broad, varied segment of the population, native-born and immigrant alike, is concerned about the future social structure of California. The extent to which political power responds to the changing ethnic mix is critical in determining how social tensions will be played out in the coming decade. An emphasis on dominance is likely to do no more than replace one power broker with another, whereas an emphasis on cooperation may resolve future social and ethnic tensions. Recall, too that the recent out-migration from California reminds us that households can vote with their feet. The future depends on maintaining viability at all levels and for all groups.

NOTES

1. The Court argued that because immigration is a federal and not a state issue, California cannot make laws with respect to immigration.
2. Recent studies have suggested that the proportion of Hispanic voters will increase dramatically in the coming decade (DeSipio, 1996).
3. In California the passage of Proposition 209 has eliminated affirmative action for university admissions.
4. Recent reports that Asians were receiving almost one-quarter of all small business loans suggest that such programs may not necessarily benefit the groups for whom they were originally designed (*Wall Street Journal*, September 9, 1997, p. B1).

Chapter 8

THE STRAINED SOCIAL FABRIC

The California demographic landscape has changed fundamentally. California and its small, medium, and large cities are now multiethnic and multiracial; the state is no longer white and European. Nor do the changes that have taken place over the last two decades show any signs of abating: the population of California will continue to grow as a result of in-migration and the high fertility rates of the new immigrants, especially those from Mexico and Central America. Even if immigration to California were to cease immediately, the processes now under way would guarantee continued population expansion and change.

Much of the debate about immigration has focused on its costs. Even though it is clear that the economic costs of immigration sharply outweigh its benefits to many local communities, and even perhaps to whole states, ultimately the critical issue regarding immigration is not costs but the social changes that are occurring as part of the immigration process. In the coming decades California and its population must wrestle with these social changes and the complex interactions they have set in motion.

In the Preface I asked, Will the social fabric stretch or will it tear? The title of this book implies another question, Will the cauldron boil or will it simmer? Will immigration eventually produce a stable social blend or will it produce a contentious mix of competing groups marked by social isolation, balkanization, and antagonism? The previous chapters offered a window on these issues and indicated where the problems are emerging, but the answers to these questions also depend on changes in current patterns and policies. Those chapters also provided a rich set of facts and data on social trends, which are essential to understanding how California's social fabric may stretch and change in the coming decades. My tentative final assessment is that the fabric of California society can stretch, but only with considerable help from government and the stakeholders.

A DIVIDED CALIFORNIA

The research reported in the previous chapters presents a picture of a bifurcated immigrant society, a population increasingly divided between rich and poor, and a complex set of local outcomes in California. Immigrants are poor and not so poor (only a few are wealthy). Some new arrivals have already made and will continue to make social and economic progress; others are linguistically isolated. In addition, immigrants are spatially selected between north and south and between urban and agricultural communities, with very different consequences.

It is impossible to ignore the clear division between most Asian and Middle Eastern immigrants and those from Mexico and Central America. Apart from the Southeast Asian refugee immigrants and their parents, most Asians have incomes similar to the income averages for the state overall. The more recent waves of Asian immigrants do not differ much in educational levels from earlier arrivals, and they advance just as rapidly. Moreover, intermarriage is high for some Asian immigrants. As Cheng and Yang (1996) argue, the notion of Asians as a "model minority"—hardworking, upwardly mobile, and "fitting in"—is not completely far-fetched: Asians as a group outperform Hispanics as a group.

Asian immigrants do differ according to their countries of origin, especially the refugee populations from Laos, Vietnam and Cambodia. Although many new Asian immigrants maintain strong links with their native land, most second-generation Asian native-born residents have only tenuous links with their parents' country of origin. Certainly, large numbers of Asians have increased their socioeconomic status by the second generation. Certainly, too, second- and third-generation Asians are taking advantage of the California education system to further improve their status. Some Asians have succeeded as entrepreneurs; others have joined the professions. However, the picture is not all rosy: both in California as a whole and in Southern California, Asians' earnings have not yet matched those of native-born whites.

Racial differences have slowed Asian assimilation, but the emphasis they put on education and the professions is transforming both Asians and California. The future path of Asian immigration is probably related as much to the size of continuing flows as to the characteristics of the Asian populations themselves. Asians, unlike Hispanics, are geographically far removed from their countries of origin. They come from many different ethnic backgrounds, and each Asian group is much smaller than the Mexican immigrant group. Thus the likelihood of intermixing for Asians is maximized. In the manner of European immigrants of earlier decades, who were also far from "home," the Asians' greater distance from "home" makes them more likely to identify with their new country. This point is supported by Asians' high naturalization rates and rapid acquisition of the English language. In contrast, Mexican Americans have less incentive to assimilate because they can easily maintain their ties to Mexico and even exist in a "Mexican world" within the greater world of the United States (Rodriguez, 1996).

Middle Eastern immigrants are also succeeding. Like the Asians, the Middle Eastern immigrants come from 20 or 30 different cultural backgrounds. They include Armenians, Israelis, Iranians, and Arabs. Many of these immigrants are professionals and highly skilled. They possess human as well as financial capital. In general, their households are small and their per capita incomes are high (Bozorgmehr, Der-Martirosian, and Sabagh, 1996). Large numbers of Middle Eastern immigrants are a recent phenomenon, but the successive waves of immigrants benefit from approximately the same levels of human and economic capital. Recent immigrants are doing as well as earlier waves.

In contrast, Latino immigrants, who are providing much of the low-cost unskilled labor that is helping to support California's continued economic expansion, are starting at the bottom. The evidence shows that their progress is marginal at best. Later waves of immigrants are doing much less well in relative terms than earlier waves. Although some Latinos are joining the middle class, they are only slowly closing the gap with native-born whites, or indeed with other immigrant groups. Economically, they are similar to the native-born black population, with whom they compete for jobs and public assistance. Birth rates are high, family reunification continues to increase the size of the immigrant pool, and an increasing proportion of Latino immigrants and their children are living in poverty.

The problem for Latino immigrants is compounded by their low levels of human capital when they arrive and by only modest educational gains for their native-born children. Their links to Mexico (Mexicans in the United States send $3 billion a year in remittances to their families in Mexico), their low level of naturalization, and the low level of their occupations will continue to limit the progress of Mexicans up the socioeconomic ladder. They mow lawns, care for children, work in garment sweatshops, and keep the service economy running. Ironically, however, their future will determine the future of the state's social structure.

Just as California's immigrants are divided, so is the immigrant geography. The Northern California counties in the Bay Area are unlike the sprawling metropolitan counties of Southern California. To a great extent, but not entirely, this difference is created by the difference in the immigrant composition of the two regions. The south is dominated by Latino immigrants, the north by Asian immigrants. Still, Mexican immigrants account for only about half of all the immigrants to Los Angeles County in the south, and they also contribute a sizable proportion of immigrants to Alameda and Santa Clara Counties in the north.

The difference in composition leads to differences in human capital, in occupations, and in economic outcomes. The immigrants to Northern California are more likely to follow professional occupations, to earn higher incomes, to be homeowners, and to be naturalized. It is not unreasonable to argue that they are more fully assimilated than immigrants in Southern California.

POPULATION GROWTH AND THE FUTURE

The new immigrants are beginning to assimilate, but what will happen to the assimilation process in the face of large increases in population? Just how many immigrants can California accommodate? It is quite possible that the very large numbers of immigrants who are arriving in Southern California will overwhelm the state's metropolitan areas by the sheer size of their flows, and that consequently they will be unable to follow the path of earlier immigrants—not because their aspirations are lower, but because their numbers are too great to be absorbed by the changing job markets of the U.S. economy. Thus we must ask, How many people can the state continue to absorb?

The most recent population projections by the Census Bureau suggest that California's population will be around 50 million in 2025. A very large proportion of the state's current residents will still be alive in 2025— less than 30 years away. Natural increase due to the children of immigrants who are already here, a continuing stream of additional new immigrants, and possible migration from other states will increase the population by more than 18 million over the next 25 years. Most of the increase will begin in the new century, when the new young citizen children of the current large waves of immigrants begin to have children of their own. To put this increase in perspective, it is useful to recall that California's population was just under 20 million in 1970, before the rapid buildup in immigration, and jumped to 29.8 million in 1990, after 2 decades of very rapid immigration. Today the population is approaching 33 million.

Census data and the figures in Chapter 3 make it very clear that much of the growth in California's population will derive from births to people who are already in the state. Moreover, the growth due to fertility will occur mainly in the Latino community. The other source of new growth, perhaps as much as 40%, will be from continued legal and illegal immigration. We know that about one-third of all immigrants who come to the United States choose to live in California; therefore, it is entirely possible that the year 2025 population estimate is too low. This estimate is based on a continuing outflow of migrants from California and on the fact that economic changes can affect the rate, and hence the number, of people who leave the state. It is clear and incontrovertible that projected growth will create very large youthful and elderly populations at the same time. Census Bureau estimates suggest that one-third of the population will be under 20 years old in 2025; at the same time, more than 13% of the population will be over 65. This anomalous, bifurcated population will require schools and old-age retirement homes simultaneously. Moreover, the young population will be mostly Asian and Latino, while the aging population will be mostly non-Hispanic white.

The numbers are not the only issue: the impact of a rapidly growing population is also a concern. How will the dramatic increase in population intersect with California's infrastructure, especially its educational infrastructure? Will the new ethnic populations cooperate or be in conflict?

GROWTH AND THE INFRASTRUCTURE

For those who see immigration in a positive light, adding another 20 million people will merely create more extensive metropolitan environments. Optimists predict that as jobs and businesses disperse from the central city to the suburbs and beyond, travel times and air pollution in California will decrease. As the state shifts into a 21st century mentality of growth and expansion, home values will increase, tax revenue will grow, and an increasing youthful population will create opportunities in a wide variety of apparel and youth-oriented industries. Wisely managed growth will increase the quality of life for all, both native-born and immigrants alike. As demographers point out, numbers per se are not the issue: "If 17 million more people come to the state, that's one thing. If none of them recycles, that makes it a bigger problem" (*Los Angeles Times*, 25 August 1997, B1).

A more pessimistic view revolves around the question, Are Californians politically aligned, and if so, for how long, to accept the reality that the state is absorbing and educating the majority of its immigrants without full federal reimbursement? Many of the new immigrants come to California from nations where education is less of a priority than in this country. Are Californians willing to pay to educate these poorly educated new arrivals to create the necessary human capital to fit the immigrants into the U.S. economy of the 21st century?

The education issue has just emerged full-grown in the national consciousness; it is clear that a crisis in classroom space is looming. Anecdotal evidence suggests that there will not be enough classrooms or enough teachers to deal with the coming surge of children born to the new immigrants. At the moment, California is doing a particularly poor job of educating the Latino immigrants.

THE FUTURE AND EDUCATION

An emerging consensus has it that educational institutions are more important than either labor markers or the social welfare system for the well-being of the community (Reitz, 1998), and that education is at the heart of the changing social and economic structure. Without significant ongoing investment in education, society will not gain the benefits of increased human capital and greater industrial productivity. Reitz (1998) points out that native-born American whites have accumulated massive human capital; this is the favorable situation that we observed earlier regarding the comparative-skills advantage of native-born whites over recent immigrants.

The data I presented in previous chapters document a rapid increase in poor, less-well-educated parents. Theory tells us that children with poor and poorly educated parents are disadvantaged themselves. Thus a dual need exists: to enable new immigrants to improve their own economic well-being and to provide their children with a solid education. We know that immigrants will continue to move to places where other immigrants live already. Thus the crises of creating an educated population will be most severe in these locations that are already strained. Public education provided the previous flows of migrants with the means to adjust to their new society; it spurred upward mobility for their children if not for the immigrants themselves. Today's evidence suggests that the school systems in the large immigrant cities are unequal to the complex demands of educating this large new wave of citizen children of immigrants (Rolph, 1992). Moreover, the politically correct celebration of cultural and linguistic diversity may be slowing rather than enhancing the educational progress of the poorest and least-well-equipped immigrants.

Despite the fact that California's increasing youthful population requires more money for its education, the state's increase in spending for education is relatively small—indeed, less than the average growth in state spending as a whole. Spending on health, welfare, and corrections is almost two times greater than the spending on K–12 education (California Legislative Analysts Office, 1995). The need for increased spending on education is generally acknowledged, but we need to commit ourselves to far more than a simple increase in spending. Only a fundamental national rethinking of the whole educational process will set the immigrants already in California on the path to increasing their human capital and eventually achieving equality with the native-born population. As others have noted, we ignore the new immigrants' educational needs at our own peril.[1]

I have paid a good deal of attention to the issue of education because in the long run it is here that immigration and social change will intersect. The old style of immersion teaching of the early 20th century is not an option in the late 20th and early 21st centuries. But neither is the continuing low levels of attention to education that have characterized California in the last decade and a half. California is currently ranked 41st among the 50 states in spending on education.[2] The state cannot spend so little on developing human capital and at the same time integrate the new immigrant children into its 21st-century society. But California, even with large budget increases, cannot assume the whole burden of educating the new immigrants. It is clear from a California perspective that there is a critical need for greater federal recognition of the special role of California as the immigrant entry port.

IMMIGRATION AND THE ENVIRONMENT

Until recently, the impact of immigration on the environment was discussed little, if at all. It is still much less of an issue than the impact of immigration on the economy or our societal organization. It has been suggested that large-scale immigration will increase consumption, thereby causing further harm to the environment. Impacts on the environment, however, are still viewed as an extension of the larger issues of population growth and high fertility rather than of immigration per se. There is no research to provide any measures of the environmental impacts of immigration. The consensus is still closer to the recent National Research Council (1997) report that the concern over immigration is related to population growth and to the impact on the infrastructure as a whole, rather than to any impact on the physical environment. That situation may change in the future.

Yet the issues concern more than education and the environment; they relate to providing housing, freeways, health services, and all the infrastructure of a modern society. How these services are (or are not) provided is critical in determining how well our society as a whole will serve its constituent parts. Success or failure in this regard will directly affect the critical issue of conflict and its resolution.

THE ARGUMENTS FOR SKILL-BASED IMMIGRATION

The need to provide substantial education for the immigrants and their children who have already arrived in California raises a more general issue: What course is to be followed in a well-crafted immigration policy? As the research presented in this book suggests, and as others (e.g., Mongia, 1997) have already noted, the United States needs a sounder, more carefully articulated immigration policy. A clear policy with appropriate checks and balances could fulfill our humanitarian commitment to immigration without causing serious disadvantageous to our native-born population. Even the most ardent proponents of immigration are beginning to recognize that the country cannot take in *all* of the world's displaced people. A rational policy would emphasize skill and education over family reunification, and in most circumstances would require an age limit on immigrants. Although it may seem humanitarian to allow those immigrants already here to bring in their elderly parents, it should be recognized that family reunification also works in the opposite direction: those who want to join elderly relatives can do so in their native countries. Alternatively, a limit on eligibility for benefits can be related to lifetime work in the United States. Who should be held financially responsible for the elderly relatives of immigrants to the United States, the U.S. taxpayer or the immigrant adult children who want to bring their parents to the United States? Certainly, adding the elderly relatives of recent immigrants to our population will only exacerbate our society's existing problems regarding adequately supporting the native-born elderly population.

The evidence I presented in the chapters on education and earnings is congruent with much other research, which recognizes at last that the very large number of immigrants who have not completed high school are competing both with less-skilled native-born residents and with other recent immigrants. This may be good news for those who employ low-skill labor, but the long-term consequences of such a strategy are likely to include the increasing marginalization and social exclusion of particular immigrant groups.

Skill-based immigration would slow the growth of the low-end labor force, thus providing some protection for both low-skill native-born residents and the new immigrants already in the country. The high school drop-out rate among native-born Hispanics is extremely high, and many of these young adults will never acquire the skills to enter a high-technology labor force. Currently, agriculture may be suffering a shortage of low-skill labor, but it is very unlikely that low-skill entry-level labor will be lacking in the long run. This is yet another reason to change the basis of immigration and to concentrate our efforts on the almost 4 million new immigrants who have already entered California in the past decade.

CONFLICT AND COOPERATION

Emigration and immigration, by definition, involve the intersection of nations. Migrants move between different countries, and this movement involves different cultures and their differing social and political organizations. Immigrants come with different customs, different expectations, and different views of the world. Some come from democratic societies, others from countries marked by political repression and limited rights. The large-scale flows of people into the United States in the last decade of the 20th century have been intensely political. Such flows can destabilize nations and change their notions of citizenship and national identity.

The changes in the former Soviet Union are an obvious recent example. These changes emerged from a fundamental shift in the concept of civil rights, which led to a greater emphasis on international humanitarian rights and a greater openness in American foreign policy. In turn, the new openness in world politics has had dramatic consequences for the ability of nation-states to control their borders (Hollifield, 1996, p. 62). The increasing worldwide flows of people beginning with the guest worker programs in the 1950s and 1960s (the bracero program in the United States can be seen in this same context) were the beginning of ethnic transitions, first in Western Europe, and then in the United States. These flows triggered issues of citizenship, issues of ethnic rights, and the numerous issues surrounding multiculturalism. In the microcosm of California, the largest state in the United States, the conflicts or cooperation surrounding multiculturalism are important to the future of ethnic assimilation.

What are the potential trajectories of ethnic relations in California? Does the future hold conflict or cooperation for the increasingly diverse groups entering the state? What is the nature of the cauldron and the nature of assimilation? To recall Rodney King's phrase after the 1992 Los Angeles riots, can we all "get along"? Will the future revisit the conflict of the past, or will accommodation prevail? In some sense, we now return to the issues of diversity and assimilation that I raised in the opening chapter of this book.

Conflict takes many forms, from disagreement generated by prejudice to tensions arising from conflict over jobs and political power. At the extreme, the 1992 riots were a volatile manifestation of minority/majority conflict and of interethnic conflict. Race and ethnic relations are much more complicated today than they were in the 1960s and 1970s, when the debates revolved around outlawing segregation and creating equal opportunities for blacks. Today's conflict has four subdimensions: cultural differences, racial and interethnic tension, economic conflict, and political power struggles. Each of these is intertwined with recent immigration.

Cultural differences, cultural misunderstanding, and cultural intolerance are the background to tensions that arise in response to imagined and/or real cultural slights. When large numbers of new immigrants are mixed together in confined urban spaces, cultural differences will be highlighted rather than minimized, and are then more likely to lead to tension and conflict. Abelmann and Lie (1995) document the tensions between the Korean and black communities in Central Los Angeles. Korean store owners defending their possessions with guns and the conflicts arising between Korean store owners and their black teenager customers are not simply a case of black–Korean distrust. They represent the problems of new immigrants flooding into older inner-city neighborhoods and making for themselves what the earlier residents perceive as successful entrepreneurial lives—even if such "success" is more a matter of perception than reality. We know, however, that many Asians have strong own-race preferences and do not want to live close to the inner-city black population (Abelmann and Lie, 1995, p. 106; Clark, 1992).

The extent of cultural conflict is probably greater than is realized. The insightful commentary by Abelmann and Lie (1995) is only one indication of the complexity of this topic and its outcomes, and it is equally applicable to other interethnic relationships. The cauldron may be created by different cultures thrust together in spatial proximity, but it bubbles with specific interethnic tensions over the willingness to live in the same neighborhoods and to attend the same school together. It boils with particular intensity over the competition for jobs and political power.

Despite long-standing attempts to emphasize minority cohesion, interethnic tensions today are as strong as tensions between whites and blacks. Indeed, in time, the tensions between blacks and whites may even be replaced by interethnic tensions. Certainly, the data on residential living preferences emphasize strong own-race preferences (Clark, 1992). Both Hispanics and Asians strongly prefer their own race and are much less tolerant of other racial and ethnic groups. In fact, of all groups, blacks are most willing to live in neighborhoods with 50/50 combinations of blacks and other races. This pattern is not replicated for the other racial/ethnic groups. Hispanics show a very strong preference for residential integration with whites, but a low preference for living alongside blacks. Asians show a very high preference for integration with whites and low preferences for living with Hispanics and blacks (Clark, 1992).

According to polls, all ethnic groups believe that direct racism is no longer a white trait (*Los Angeles Times* Poll, 1992, November 16, JJ5). Whites, blacks, Hispanics, and Asians were near unanimous (87–95%) in their belief that all groups are guilty of prejudice. All groups believed that most discrimination was directed against Hispanics and blacks. A majority of all groups thought that race relations were not improving. The interethnic tension, however, may be the most troubling in the long run, given the continuing pattern and process of immigration.

Survey data and newspaper reports provide fairly strong anecdotal evidence of the tension between blacks and Hispanics (see, e.g., *Los Angeles Times,* 21 August 1994, A33; 25 March 1995, B3). These ethnic tensions are often hidden from the metropolitan population as a whole, but they surface in conflicts in the public schools. It is increasingly common to read newspaper reports that highlight the existence of racial tension in California. For example, in 1995, helmeted police used pepper spray to subdue a fight between 40 black and Hispanic students outside Richmond High School, capping a week of tension between the two groups (*San Francisco Examiner,* 14 May 1995, p. C11). One teenager was shot and about 30 students were arrested in two racially motivated fights near Leuzinger High School in Lawndale (*Los Angeles Times,* 25 March 1995, p. B3). Although such incidents dramatize the issue of racial tension, the consequences for the educational process extend much further. Many students stay away from school, and the learning process is disrupted. In addition, the climate of fear and possible retaliation is antithetical to peaceful learning. The 3 days of education missed at Richmond High School are not easily replaced.

Blacks and Hispanics are the majority in a number of central-city districts where new Hispanic immigrants have been arriving in large numbers. In those low-income neighborhoods, the two groups are competing with one another for a very small part of the economic pie and they are also battling over the political power structure. This pattern is common to metropolitan areas outside California as well, especially in Florida; there, large numbers of Cubans have moved into formerly black neighborhoods (Mohl, 1990).

Are immigrants, especially Asian and Hispanic immigrants, creating interethnic tensions by displacing black workers? Mounting evidence suggests that this is so; certainly there is a strong perception that they are displacing native-born workers (Field Institute, 1993). Although there has been no published study, it is widely perceived that black (union) workers are gradually being displaced by Hispanic (non-union) workers in Los Angeles. This displacement has been reported on computer networks and is "perceived" as an effect of immigration in the black community. The data from the National Research Council (1997) study indicates job displacement, and the research cited in Chapter 4 suggests that immigrants are taking over certain economic niches. A study by the Urban Institute suggests that even the niches that have traditionally provided opportunities for low-skilled immigrants are likely to shrink in the future (Enchautegui, 1997).

To reiterate an earlier statement, political conflict may be the area of greatest tension in the long run. Tension exists between black and Latino political groups. In the eyes of black leaders, Latinos pushing for power are taking advantage of gains made by the struggles of African Americans in the civil rights movement, but are not acknowledging their debt to blacks. The issue, however, is larger than a dispute in particular places. It is symptomatic of one possible future for communities in Southern California, and eventually even the nation: political upheaval as groups struggle for group power rather than individual influence.

The latest discussion of "Americanization" by the U.S. Commission on Immigration Reform (1997) suggests that an emphasis on the commonality of the "principles and values embodied in the American Constitution" (p. 5) will provide a template for overcoming the potential divisions in our society. Despite the risk that such an emphasis will be regarded as a nativist view, the report is careful to spell out how diversity is compatible with national unity. At the same time, the Commission on Immigration Reform emphasizes the need for a shared commitment to a common language and ideals. Again, education is a key to the process of assimilation, and will accomplish the goals of unity and communality.

CONCLUSION

The review in this chapter brings me back full circle to the issues of assimilation versus separation and the keys to a successful immigrant society. I am less sanguine than the authors of the recently released study by the National Research Council (1997); they view immigration quite favorably, emphasizing the $10 billion that immigrants may be adding to the economy each year. That amount, however, is only about 1% of the $7 trillion U.S. economy. Moreover, a society consists of more than an economy. Even so, the evidence suggests that overall the gains at the national level outweigh the costs, but the social and economic costs at the local level mostly outweigh the local gains.

As I have stressed repeatedly, it is not the costs of immigration that are the central issue: it is the change in our social fabric. Our nation must accommodate the changing multiethnic complexity so that the State of California and the United States retain their identity as one society with many ethnic backgrounds and identities, rather than divide into separate ethnic and racial societies sharing a common territory but holding only weak loyalties to the nation as a whole.

The questions of changing immigration policy are tied closely to the politics of United States–Mexico relations and to the politics of race in the United States. The research reported in this book establishes the closeness of the links between California and Mexico. Almost half of all Latinos in

California are foreign-born, mostly in Mexico. But historically, discussions of race have been dominated by a biracial (black/white) view of the United States. That world has vanished, certainly for all of the metropolitan areas in the west and on the West Coast, and increasingly in the Midwest. Only certain older industrial cities on the East Coast can still see the world in black and white; and even there the situation is changing rapidly.

This book provides a rich set of graphs and tables that can be used to inform public policy and to generate conclusions about immigration in the 21st century. This material enables the reader to decide for himself or herself about the validity of my positions. The conclusions that I draw from the research suggest a modified view of immigration, in which the United States returns to a skill-based immigration policy and substantially increases and federalizes spending on education. Another recent analysis has reached the same conclusion for Southern California: the pace of immigration must be slowed if social integration is to occur and if those who are already here are to receive opportunities for socioeconomic advancement (Allen and Turner, 1997). We cannot expect the cauldron to stop bubbling, but by rationalizing immigration policy to emphasize skills for admittance and education for human capital formation for those already here, we would go a long way in reducing the tensions and decreasing the separation that currently exists.

NOTES

1. Similar arguments are presented in McCarthy and Vernez (1997).
2. California's expenditure of $4977 per pupil in 1995–96 is below the national average of $6103, and half that of New Jersey, $9967. At the same time, California ranked 15th in per capita personal income and teacher's salaries were 15% above the National average (EdSource. 1997. *EdFact—California's Rankings*. Palo Alto: California).

APPENDIX

COUNTIES & MAJOR CITIES   REGIONS

BIBLIOGRAPHY

Abelmann, N., and Lie, J. 1995. *Blue Dreams*. Cambridge, MA: Harvard University Press.

Allen, J., and Turner, E. 1989. The most ethnically diverse urban places in the United States. *Urban Geography* 10: 523–539.

Allen, J., and Turner, E. 1996. Spatial patterns of immigrant assimilation. *Professional Geographer* 48: 140–155.

Allen, J., and Turner, E. 1997. *The Ethnic Quilt: Population Diversity in Southern California*. California State University, Northridge: Department of Geography, Center for Geographical Studies.

Arroyo, J., De Leon, A., and Valenzuela, B. 1990. Patterns of migration and regional development in the State of Jalisco, Mexico. (Cited in J. Navarro, *The Economic Effects of Emigration: Mexico,* Chapter 6 in B. Asch. 1994. *Emigration and The Effects on the Sending Country.* Santa Monica, CA: The Rand Corporation.

Bailey, T. 1987. *Immigrant and Native Workers: Contrasts and Competition*. Boulder, CO: Westview Press.

Baker, S. 1990. *The Cautious Welcome*. Santa Monica, CA: Rand Corporation.

Baker, K. 1992. Ramirez et al: Led by bad theory. *Bilingual Research Journal* 16: 91–104.

Baker, K., and de Kanter, A. 1983. An answer from research on bilingual education. *American Education* 56: 157–169.

Bartel, A. 1979. The migration decision: What role does job mobility play? *American Economic Review* 69: 775–786.

Bartel, A. 1989. Where do the new U.S. immigrants live? *Journal of Labor Economics* 7: 371–391.

Bayor, R. H. 1978. *Neighbors in Conflict: The Irish, Germans, Jews, and Italians of New York City, 1929–1941*. Baltimore: Johns Hopkins University Press.

Bean, F., Chapa, J., Berg, R., and Sowards, K. 1994. Educational and sociodemographic incorporation amongst Hispanic immigrants to the United States. In *Immigration and Ethnicity: America's Newest Arrivals,* B. Edmonston and J. Passel (Eds.). Washington, DC: Urban Institute Press.

Bean, F. D., Edmonston, B., and Passel, J. (Eds.). 1990. *Undocumented Migration to the United States: IRCA and the Experience of the 1980s.* Santa Monica, CA: Rand Corporation.

Bean, F., and Fix, M. 1992. The significance of recent immigration policy reform in the United States. In *Nations of Immigrants: Australia, the United States, and International Migration,* G. Freedman and J. Jupp (Eds.), pp. 41–55. Melbourne, Australia: Oxford University Press.

Bean, F., and Tienda, M. 1988. *The Hispanic Population of the United States.* New York: Russell Sage Foundation.

Berry, B. J. L., and Dahmann, D. C. 1980. Population redistribution in the United States in the 1970s. In *Population Redistribution and Public Policy,* B. J. L. Berry and L. P. Silverman (Eds.), pp. 8–49. Washington, DC: National Academy of Sciences.

Borjas, G. 1989. Economic theory and international migration. *International Migration Review* 23: 457–485.

Borjas, G. 1990. *Friends or Strangers? The Impact of Immigrants on the U.S. Economy.* New York: Basic Books.

Borjas, G. 1994. The economics of immigration. *Journal of Economic Literature* 32: 1667–1717.

Borjas, G. 1995a. Assimilation and changes in cohort quality revisited: What happened to immigrant earnings in the 1980's? *Journal of Labor Economics* 13: 201–245.

Borjas, G. 1995b. The economic benefits from immigration. *Journal of Economic Perspectives* 9 (2): 3–22.

Borjas, G., and Freeman R. (Eds.). 1992. *Immigration and the Work Force: Economic Consequences for the United States and Source Areas.* Chicago: University of Chicago Press.

Bouvier, L. 1991. *Peaceful Invasions: Immigration and Changing America.* Lanham, MD: University Press of America.

Bouvier, L. 1992. *Fifty Million Californians.* Washington, DC: Center for Immigration Studies.

Bouvier, L. 1998. Population growth and immigration, an analysis of the NRC Study. *Immigration Review* 31: 1–5.

Boyd, M. 1989. Family and personal networks in migration. *International Migration Review* 23: 638–670.

Bozorgmehr, M., Der-Martirosian, E., and Sabagh, C. 1996. Middle Easterners: A new kind of immigrant. In *Ethnic Los Angeles,* R. Waldinger and M. Bozorgmehr (Eds.), pp. 345–378. New York: Russell Sage Foundation.

Brimelow, P. 1994. *Alien Nation: Common Sense about America's Immigration Disaster.* New York: Random House.

Buenker, J., and Ratner, L. (Eds.). 1992. *Multiculturalism in the United States: A Comprehensive Guide to Acculturation and Ethnicity.* New York: Greenwood Press.

Burke, M. 1995. Mexican immigrants shape California's fertility future. *Population Today* 23 (9): 1–3.

Bustamante, J., Reynolds, E., and Hinojosa-Ojeda, R. (Eds.). 1992. *U.S.–Mexico Relations: Labor Market Interdependence.* Stanford, CA: Stanford University Press.

Butcher, K., and Card, D. 1991. Immigration and wages: Evidence from the 1980s. *American Economic Review* 81: 292–296.

Butcher, K., and Piehl, A. 1997. Immigration and the wages and employment of U.S.-born workers in New Jersey. In *Keys to Successful Immigration*, T. Espenshade (Ed.), pp. 35–54. Washington, DC: Urban Institute Press.

California Policy Seminar. 1994. *Conference Proceedings.* California Immigration—1994, Sacramento, CA, April 29. Sacramento: California State University.

California Legislative Analysts Office. 1995. *State Expenditures January 1995.* Sacramento, CA.

California Senate Office of Research. 1995. *Teen Pregnancy and Parenting in California: Background.* Sacramento, CA: Senate Office of Research.

Callis, R. 1997. *Moving to America, Moving to Home Ownership.* Washington, DC: U.S. Bureau of the Census.

Camarota, S. 1997. Five million illegal immigrants: An analysis of new INS numbers. *Immigration Review* 28: 1–4.

Camarota, S. 1998. *The Wages of Immigration: The Effect on the Low-Skilled Labor Market.* Washington, DC: Center for Immigration Studies, No. 12.

Castles, S. 1986. The guest-worker in Western Europe—An obituary. *International Migration Review* 23: 761–778.

Castles, S., and Miller, M. 1993. *The Age of Migration: International Population Movements in the Modern World.* New York: The Guilford Press.

Center for Immigration Studies. 1996. *U.S. Immigration Hot Spots.* Washington, DC: Center for Immigration Studies.

Cervantes, N., Khokha, S., and Murray, B. 1995. Hate unleashed: Los Angeles and the aftermath of Proposition 187. *Chicano-Latino Law Review* 17: 1–23.

Champion, A. 1994. International migration and demographic change in the developed world. *Urban Studies* 31: 653–677.

Cheng, L., and Yang, P. 1996. Asians: The "model minority" deconstructed. In *Ethnic Los Angeles*, R. Waldinger and M. Bozorgmehr (Eds.), pp. 305–344. New York: Russell Sage Foundation.

Chiswick, B. 1978. The effect of Americanization on the earnings of foreign-born men. *Journal of Political Economy* 86: 897–921.

Chiswick, B. 1986. Is the new immigration less skilled than the old? *Journal of Labor Economics* 4 (2): 168–192.

Chiswick, B., and Miller, P. 1995. The endogeneity between language and earnings: International analyses. *Journal of Labor Economics* 13: 246–288.

Chiswick, B., and Miller, P. 1996. The languages of the United States: What is spoken and what it means. *Read perspectives* 3: 5–41.

Chiswick, B., and Sullivan, T. 1994. The new immigrants. In *State of the Union*, Reynolds Farley (Ed.), pp. 211–270. New York: Russell Sage Foundation.

Clark, R. 1994. *The Costs of Providing Public Assistance and Education to Immigrants.* Washington, DC: Urban Institute Press.

Clark, R., Passel, J., Zimmerman, W., and Fix, M. 1994. *Fiscal Impacts of Undocumented Aliens: Selected Estimates for Seven States.* Report to the Office of Management and Budget and the Department of Justice. Washington, DC: Urban Institute Press.

Clark, W. A. V. 1992. Residential preferences and residential choices in a multi-ethnic context. *Demography* 29: 451–466.

Clark, W. A. V. 1996a. Residential patterns, avoidance, assimilation and succession. In *Ethnic Los Angeles,* R. Waldinger and M. Bozorgmehr (Eds.), pp. 109–138. New York: Russell Sage Foundation.

Clark, W. A. V. 1996b. Scale effects in international migration to the United States. *Regional Studies* 30: 589–600.

Clark, W. A. V. 1998a. Mass migration and local outcomes: Is international migration to the United States creating a new urban under class? *Urban Studies* 35: 371–383.

Clark, W. A. V. 1998b. Large scale immigration and political response: Popular reaction in California. *International Journal of Population Geography* 4: 1–10.

Clark, W. A. V., and Dieleman, F. 1996. *Households and Housing.* New Brunswick, NJ: Center for Urban Policy Research, Rutgers University.

Clark, W. A. V., and McNicholas, M. 1995. Re-examining economic and social polarization in a multi-ethnic metropolitan area: The case of Los Angeles. *Area* 28: 56–63.

Clark, W. A. V., and Morrison, P. 1991. Demographic paradoxes in the Los Angeles voting rights case. *Evaluation Review* 20: 712–726.

Clark, W. A. V., and Morrison, P. 1995. Demographic foundations of political empowerment in multi-minority cities. *Demography* 32: 183–201.

Clark, W. A. V., and Mueller, M. 1988. Hispanic relocation and spatial assimilation: A case study. *Social Science Quarterly* 69: 468–475.

Clark, W. A. V., and Schultz, F. 1997. Evaluating the local impacts of recent immigration to California: Realism versus racism. *Population Research and Policy Review* 16: 475–491.

Clark, W. A. V., and Ware, J. 1997. Trends in residential integration by socio-economic status in Southern California. *Urban Affairs Review* 32: 825–843.

Coiro, M. J., Zill, M., and Bloom, B. 1994. *Health of Our Nation's Children.* Hyattsville, MD: National Center for Health Statistics, U.S. Department of Health and Human Services.

Conner, W. (Ed.). 1985. *Mexican Americans in Comparative Perspective.* Washington, DC: Urban Institute Press.

Cornelius, W., Martin, P., and Hollifield, J. (Eds.). 1994. *Controlling Immigration: A Global Perspective.* Stanford, CA: Stanford University Press.

Corwin, A. 1973. Causes of Mexican emigration to the United States. *Perspectives in American History* 7: 557–635.

Cose, E. 1992. *A Nation of Strangers: Prejudice, Politics, and the Populating of America.* New York: Morrow.

Currie, J., and Thomas, D. 1997. *Does Head Start Help Hispanic Children?* Santa Monica, CA: Rand Corporation.

Current Population Survey, CPS. March 1994. *Machine-Readable Data File.* Washington, DC: U.S. Bureau of the Census.

Current Population Survey, CPS. March 1996. *Machine-Readable Data File.* Washington, DC: U.S. Bureau of the Census.

DaVanzo, J. 1976. *Why Families Move: A Model of the Geographic Mobility of Married Couples.* Santa Monica, CA: Rand Corporation.

DaVanzo, J. 1981. Microeconomic approaches to studying migration decisions. In

Migration Decision Making: Multidisciplinary Approaches to Microlevel Studies in Developing and Developed Countries, G. F. DeJong and R. W. Gardner (Eds.), pp. 90–129. New York: Pergamon Press.

DaVanzo, J., Hawes-Dawson, J., Valdez, R., and Vernez, G. 1994. *Surveying Immigrant Communities*. Santa Monica, CA: Rand Corporation.

Davidson, C. (Ed.). 1984. *Minority Vote Dilution*. Washington, DC: Howard University Press.

Davis, C., Haub, C., and Willete, J. 1983. *U.S. Hispanics Changing the Face of America*. Population Bulletin 38. Washington, DC: Population Reference Bureau.

De Jong, G., and Fawcett, J. T. 1981. Motivations for migration: An assessment and value-expectancy research model. In *Migration Decision Making: Multidisciplinary Approaches to Microlevel Studies in Developed and Developing Countries*, G. DeJong and R. Gardner (Eds.), pp. 186–224. New York: Pergamon Press.

De la Garza, R. 1992. *Latino Voices: Mexican, Puerto Rican, and Cuban Perspectives on American Politics*. Boulder, CO: Westview Press.

Demographic Research Unit, State of California. 1996. *Refugee Immigration to California, 1990–1994: A Summary in Tables*. Sacramento, CA.

Demographic Research Unit, State of California. 1997a. *Legal Immigration to California: in Fiscal Year 1984–1994. A Summary.* Sacramento, CA.

Demographic Research Unit, State of California. 1997b. *Legal Immigration to California in Fiscal Year 1995.* Sacramento, CA: Department of Education.

Denton, N., and Massey, D. 1988. Residential segregation of blacks, Hispanics, and Asians by socioeconomic status and generation. *Social Science Quarterly* 69: 797–817.

Denton, N., and Massey, D. 1991. Patterns of neighborhood transition in a multiethnic world: U.S. metropolitan areas, 1970–1980. *Demography* 28: 41–63.

Department of Education, State of California, *Language Census Report for California Schools, 1995.* Sacramento, CA.

Department of Health Services, State of California. 1995. *California Teen and Unwed Pregnancy Statistics.* Sacramento, CA.

Department of Health Services, State of California. 1997. *Medical Care Statistics Section.* Sacramento, CA.

DeSipio, L. 1996. *Counting on the Latino Vote: Latinos as a New Electorate.* Charlottesville: University of Virginia Press.

De-Vita, C. 1996. The United States at mid-decade. *Population Bulletin* 50 (4): 1–48.

Dieleman, F. 1993. Multicultural Holland: Myth or reality? In *Mass Migration in Europe*, R. King (Ed.), pp. 118–135. London: Belhaven Press.

Donato, K. 1994. U.S. policy and Mexican migration to the United States, 1942–92. *Social Science Quarterly* 75: 705–729.

Donato, K., Durand, J., and Massey, D. 1992. Changing conditions in the U.S. labor market: Effects of the Immigration Reform and Control Act of 1986. *Population Research and Policy Review* 11: 93–195.

Duncan, G., Brooks-Gunn, J., and Klebanov, P. 1994. Economic deprivation and early-childhood development. *Child Development* 65: 296–318.

Duncan, G., and Yeung, W. 1995. Extent and consequence of welfare dependence among America's children. *Children and Youth Services Review* 17: 157–182.

Duncan, G., Yeung, W., and Brooks-Gunn, J. 1996, May 9. *Does Childhood Poverty Affect the Life Chances of Children?* Paper presented at the Population Association of America, New Orleans.

Duncan, O., and Lieberson, S. 1959. Ethnic segregation and assimilation. *American Journal of Sociology* 64: 367–374.

Duran, B. J., and Weffer, R. 1992. Immigrants' aspirations, high school process, and academic outcomes. *American Educational Research Journal* 29: 163–181.

Edmonston, B., and Passel, J. (Eds.). 1994. *Immigration and Ethnicity: The Interpretation of America's Newest Arrivals*. Washington, DC: Urban Institute Press.

Education Demographics Unit, California Department of Education. 1995. *Language Census Report*. Sacramento, CA.

Ellis, M., and Wright, R. 1998. When immigrant are not migrants: Counting arrivals of the foreign born using the U.S. Census. *International Migration Review* 32: 127–144.

Enchautegui, M. 1997. *Low Skilled Immigrants in the Changing Labor Market*. Washington, DC: Urban Institute Press.

Espenshade, T. J. 1995. Unauthorized immigration to the United States. *Annual Review of Sociology* 21: 195–216.

Espenshade, T. J. 1997. *Keys to Successful Immigration: Implications of the New Jersey Experience*. Washington, DC: Urban Institute Press.

Espenshade, T. J., and Hempstead, K. 1996. Contemporary American attitudes toward U.S. immigration. *International Migration Review* 30 (2): 535–570.

Evans, M. D. 1986. Language skill, language usage and opportunity: Immigrants in the Australian labor market. *Sociology* 21: 253–274.

Fainstein, S., Gordon, I., and Harloe, M. 1992. *Divided Cities*. Oxford, UK: Blackwell.

Fannie Mae. 1995. *Fannie Mae National Housing Survey, 1995: Immigrants, Homeownership, and the American Dream*. Washington, DC: Fannie Mae Corporation.

Farley, R., and Frey, W. 1994. Changes in the segregation of whites from blacks during the 1980s: Small steps toward a more integrated society. *American Sociological Review* 59: 23–45.

Field Institute. 1993. *California Poll*. San Francisco, California.

Fishman, J. 1972. *The Sociology of Language*. Rowley, MA: Newbury House.

Fix, M., and Zimmerman, W. 1994. After arrival: An overview of federal immigrant policy in the United States. In *Immigration and Ethnicity: America's Newest Arrivals*, B. Edmonston and J. Passel (Eds.), pp. 251–286. Washington, DC: Urban Institute Press.

Frey, W. 1995. The new geography of population shifts: Trends towards balkanization. In *State of the Union, America in the 1990s, Vol. 2: Social Trends*, R. Farley (Ed.), pp. 271–334. New York: Russell Sage Foundation.

Frey, W. 1996a. Immigration, domestic migration, and demographic balkanization in America: New evidence for the 1990s. *Population and Development Review* 22: 741–763.

Frey, W. 1996b. Immigrant and native migrant magnets. *American Demographics* November, 1–5.

Frey, W. 1996–1997. Immigration and the changing geography of poverty. *Focus* 18 (2): 24–28.

Frey, W., and Liaw, K. 1996. *The Impact of Recent Immigration on Population Redistribution within the United States.* Ann Arbor, MI: Population Studies Center, University of Michigan, Report No. 96-376.

Friedberg, R., and Hunt, J. 1995. The impact of immigrants on host country wages, employment, and growth. *Journal of Economic Perspectives* 9 (2): 23–44.

Gallup Poll Organization. 1993, July. *Monthly Poll*, No. 334.

Garcia, R. J. 1995. Critical race theory and Proposition 187: The racial politics of immigration law. *Chicano-Latino Law Review* 17: 118–148.

Garcia y Griego, M. 1994. History of U.S. immigration policy. In *California Policy Seminar, Conference Proceedings. California Immigration—1994, Sacramento, California, 29 March.* Sacramento, CA: California State University Press.

Garvey, D. L. 1997. Immigrants' earnings and labor-market assimilation: A case study of New Jersey. In *Keys to Successful Immigration: Implications of the New Jersey Experience,* T. J. Espenshade (Ed.), pp. 291–336. Washington, DC: Urban Institute Press.

Glazer, N. 1988. *The New Immigration: A Challenge to American Society.* San Diego: State University of San Diego Press.

Glazer, N. 1993. Is assimilation dead? *Annals of the American Association of Political and Social Science* 510: 122–136.

Gordon, M. 1964. *Assimilation in American Life.* New York: Oxford University Press.

Greenwood, M. 1983. Regional economic aspects of immigrant location patterns in the United States. In *U.S. Immigration and Refugee Policy,* M. Kritz (Ed.), pp. 233–248. Lexington, MA: Lexington Books.

Hakuta, K. 1986. *Mirror of Language: The Debate on Bilingualism.* New York: Basic Books.

Hamnett, C. 1994. Social polarization in global cities: Theory and evidence. *Urban Studies* 31: 401–424.

Hansen, N. 1990. Do producer services induce regional economic development? *Journal of Regional Science* 30: 465–476.

Haub, C. 1997. *Statistics on Mexican Fertility.* Washington, DC: Population Reference Bureau.

Haveman, R., and Wolfe, B. 1995. The determinants of children's attainments: A review of methods and findings. *Journal of Economic Literature* 33: 1829–1878.

Heer, D. M. 1990. *Undocumented Mexicans in the United States.* Cambridge, UK: Cambridge University Press.

Heer, D. M. 1996. *Immigration in America's Future: Social Science Findings and the Policy Debate.* Boulder, CO: Westview Press.

Heer, D. 1998. The legal status of the children of undocumented Mexican immigrants in Los Angeles County. *Proceedings of the Social Statistics Section of the American Statistical Association.*

Heim, M., and Austin, N. 1995. *Fertility of Immigrant Women in California.* Paper presented at the annual meeting of the Population Association of America, April 6–8, San Francisco.

Henderson, J., and Ionnides, Y. 1985. Tenure choice and the demand for housing. *Economica* 53: 231–246.

Heskin, A. 1983. *Tenants and the American Dream.* New York: Praeger.

Hirsch, E. D. 1987. *Cultural Literacy: What Every American Needs to Know.* Boston: Houghton Mifflin.

Hoefer, M. D. 1989. *Characteristics of Aliens Legalizing under IRCA.* Baltimore: Population Association of America.

Hoefer, M. D. 1991. *The Legalization Program in California.* Paper presented at the Fourth Annual Demographic Workshop, Los Angeles, April 15.

Hoffman, E. 1989. *Lost in Translation: A Life in a New Language.* New York: Dutton.

Holder v. Hall. 1994. 114 S. Ct. 2581, 129 L. Ed 2d 687.

Holland, R. 1986. *Bilingual Education: Recent Evaluation of Local School District Programs and Related Research on Second Language Learning.* Washington, DC: Congressional Research Service.

Hollifield, J. 1996. The political economy of immigration: Markets versus right in Europe and the United States. *Schriften des Zentralinstituts für Frankishe Landeskunde und Allegemeine regionalforschung an der Universität Erlungen-Nürnberg, Wirkungen van Migration auf Aufnehmend. Gesellschaften,* pp. 59–86.

Huddle, D. 1993. *The Costs of Immigration.* Washington, DC: Carrying Capacity Network.

Hugo, G. 1981. Village–community ties, village norms, and ethnic and social networks: A review of evidence from the third world. In *Migration Decision Making: Multidisciplinary Approaches to Microlevel Studies in Developed and Developing Countries,* G. DeJong and R. Gardner (Eds.), pp. 186–224. New York: Pergamon Press.

Immigration and Naturalization Service. 1995. *Statistical Yearbook.* Washington, DC: Superintendent of Government Documents.

Immigration and Naturalization Service. 1996. *Statistical Yearbook.* Washington, DC: Superintendent of Government Documents.

Isserman, A. 1993. United States immigration policy and the industrial heartland: Laws, origins, settlement patterns, and economic consequences. *Urban Studies* 30 (2): 237–265.

Jaeger, D. 1995. *Skill Differences and the Effects of Immigrants on the Wages of Natives.* White Paper 273. Washington, DC: Office of Employment, Bureau of Labor Statistics, U.S. Department of Labor.

James, F. J. 1995. Urban economies: Trends, forces, and implications for the president's national urban policy. *Cityscape: A Journal of Policy Development and Research* 1: 67–123.

Jargowsky, P. 1996. *Poverty and Race: Ghettos, Barrios, and the American City.* New York: Russell Sage Foundation.

Johnson, K., and Williams, M. 1981. *Illegal Aliens in the Western Hemisphere: Political and Economic Factors.* New York: Praeger.

Joint Economic Committee. 1997a. *Economic Report of the President.* Washington, DC: U.S. Government Printing Office.

Jones, R. 1982. Undocumented migration from Mexico: Some geographical questions. *Annals of the Association of American Geographers* 72: 77–87.

Juhn, C., Murphy, K., and Pierce, B. 1993. Wage inequality and the rise in returns to skill. *Journal of Political Economy* 101 (3): 410–442.

Kallen, H. 1995, 18 February. Democracy versus the melting pot. *Nation,* pp. 190–194. (Cited in Rose, P. 1993. Of every hue and caste: Race, immigration, and the perceptions of pluralism. *Annals of the American Academy of Political and Social Science* 530: 187–202.)

Kao, G., and Tienda, M. 1995. Optimism and achievement: The education and performance of immigrant youth. *Social Science Quarterly* 76: 1–19.

Kasarda, J. 1994. *Industrial Restructuring and the Consequences of Changing Job Locations.* Chapel Hill, NC: Kenan Institute of Private Enterprise, University of North Carolina.

Kersten, E. 1995. *Teen Pregnancy and Parenting in California: Background.* Sacramento, CA: Senate Office of Research, California Legislature.

Krashen, S. 1982. *Principles and Practice in Second Language Acquisition.* New York: Oxford University Press.

Kritz, M., and Zlotnik, H. 1992. Global interactions: Migration systems, processes, and policies. In *International Migration Systems,* M. M. Kritz, L. Lim, and H. Zlotnik (Eds.), pp. 1–18. Oxford: Oxford University Press.

Lalonde, R., and Topel, R. 1991. Labor market adjustment of increased immigration. In *Immigration Trade and the Labor Market,* J. Abowd and R. Freeman (Eds.), pp. 167–200. Chicago: University of Chicago Press.

Lapham, S. J. 1993. *The Foreign-Born Population of the United States: 1990.* Washington, DC: Ethnic and Hispanic Branch, Bureau of Census, U.S. Department of Commerce.

Layard, R., Blanchard, O., Dornbush, R., and Krugman, P. 1992. *East-West Migration: The Alternatives.* Cambridge, MA: MIT Press.

Levy, F., and Murnane, R. 1992. U.S. earnings levels and earnings inequality: A review of recent trends and proposed explanations. *Journal of Economic Literature* 30: 1333–1381.

Lieberson, S. 1963. *Ethnic Patterns in American Cities.* New York: Free Press.

Lieberson, S., and Curry, T. J. 1971. Language shifts in the United States: Some demographic clues. *International Migration Review* 5: 125–137.

Lieberson, S., and Waters, M. C. 1988. *From Many Strands: Ethnic and Racial Groups in Contemporary America.* New York: Russell Sage Foundation.

Light, I., and Bhachu, P. (Eds.). 1993. *Immigration and Entrepreneurship.* New Brunswick, NJ: Transaction Books.

Light, I., and Roach, E. 1996. Self-employment: Mobility ladder or economic life boat? In *Ethnic Los Angeles,* R. Waldinger and M. Bozorgmehr (Eds.), pp. 193–214. New York: Russell Sage Foundation.

Locke, D. C. 1991. *Increasing Multicultural Understanding: A Comprehensive Model.* Newbury Park, CA: Sage.

Logan, J. R., Alba, R., and McNulty, T. 1994. Ethnic economies in metropolitan regions: Miami and beyond. *Social Sciences* 72: 691–724.

Lopez, D. 1978. Chicano language loyalty in an urban setting. *Sociology and Social Research* 82: 267–278.

Lopez, D. 1982. *Language Maintenance and Shift in the United States Today: The Basic Patterns and Their Implications.* Los Alamitos, CA: National Center for Bilingual Research.

Lopez, D. 1996. Language: diversity and assimilation. In *Ethnic Los Angeles,* R. Waldinger and M. Bozorgmehr (Eds.), pp. 139–163. New York: Russell Sage Foundation.

Lopez-Garza, M. 1989. Immigration and economic restructuring: The metamorphosis of Southern California. *Californian Sociologist* 12: 93–110.

Margolis, J. 1995. Closing the doors to the land of opportunity: The constitutional controversy surrounding Proposition 187. *University of Miami Inter-American Law Review* 26: 363–401.

Martin, P. 1995. *The Mexican Crisis and Mexico–U.S. Migration.* Unpublished report, University of California, Davis, Department of Economics.

Martin, P., and Midgley, E. 1994. Immigration to the United States: Journey to an uncertain destination. *Population Bulletin* 49: 1–47.

Masnick, G. S., Pitkin, J. R., and Brennan, J. 1990. Cohort housing trends in a local housing market: The case of Southern California. In *Housing Demography,* D. Meyers (Ed.), pp. 157–173. Madison: University of Wisconsin Press.

Massey, D. 1988. Economic development and international migration in comparative perspective. *Population and Development Review.* 14: 383–414.

Massey, D. 1990. Social structure, household strategies, and the cumulative causation of migration. *Population Index* 56: 3–26.

Massey, D. 1995. The new immigration and ethnicity in the United States. *Population and Development Review* 21: 631–652.

Massey, D., Arango, J., Hugo, G., Kouaouci, A., Pellegrino, A., and Taylor, J. E. 1993. Theories of international migration: A review and appraisal. *Population and Development Review* 19: 431–466.

Massey, D., and Denton, N. 1987. Trends in the residential segregation of blacks, Hispanics, and Asians, 1970–1980. *American Sociological Review* 52: 802–825.

Massey, D. S., Donato, K. M., and Liang, Z. 1990. Effects of the Immigration Reform and Control Act of 1986: Preliminary data from Mexico. In *Undocumented Migration to the United States: IRCA and the Expreience of the 1980s,* F. D. Bean, B. Edmonston, and J. Passel (Eds.), pp. 183–210. Washington, DC: Urban Institute Press.

McArdle, N. 1997. Homeownership attainment of New Jersey immigrants. In *Keys to Successful Immigration: Implications of the New Jersey Experience,* T. J. Espenshade (Ed.), pp. 337–369. Washington, DC: Urban Institute Press.

McCarthy, K., and Vernez, G. 1997. *Immigration in a Changing Economy: The California Experience.* Santa Monica, CA: Rand Corporation.

Mexico–United States Binational Study. 1998. *Migration between Mexico and the United States.* Washington, DC: U.S. Commission on Immigration Reform.

Mincer, J. 1978. Family migration decisions. *Journal of Political Economy* 86: 749–773.

Miyares, I. 1997. Changing perceptions of space and place as measures of Hmong acculturation. *Professional Geographer* 42: 214–224.

Mohl, R. 1990, July. On the edge: Blacks and Hispanics in metropolitan Miami since 1959. *Florida Historical Quarterly.*

Mongia, S. 1997, August. *Mend It, Don't End It: A Case for a Skills-Based Immigration Policy.* Report of the Jerome Levy Economics Institute of Bard College 7, No. 3.

Moore, D. 1993, July. Americans feel threatened by new immigrants. *Gallup Poll Monthly,* pp. 3–16.

Moore, J., and Pachon, H. 1985. *Hispanics in the United States.* Englewood Cliffs, NJ: Prentice-Hall.

Moreno, M. 1992. *Impact of Undocumented Persons and Other Immigrants on Costs, Revenues, and Services in Los Angeles County.* Unpublished report, Urban Research Section, Los Angeles County.

Morrison, P. 1998. Demographic influences on Latinos political empowerment: Comparative local illustrations. *Population Research and Policy Review* 17, forthcoming.

Morrison, P., and McCarthy, K. F. 1982. *Demographic Forces Reshaping Small Communities in the 1980s.* Santa Monica, CA: Rand Corporation.

Muller, T., and Espenshade, T. J. 1985. *The Fourth Wave: California's Newest Immigrants.* Washington, DC: Urban Institute Press.

Muus, P. 1996. *Diverging Development of North to South Migration in Two Migration Systems: A Comparison of International Migration from Mexico to the U.S.A. and from the Maghreb and Turkey to the European Union.* Utrecht, The Netherlands: European Research Center on Migration and Ethnic Research.

Myers, D., and Lee, S. 1995. *Immigration Cohorts and Residential Overcrowding: A Double Cohort Method for Estimating Improvement over Time in Southern California.* Paper presented at the annual meeting of the Population Association of America, San Francisco, April 20.

Myers, D., and Lee, S. W. 1998. Immigrant trajectories into homeownership: A temporal analysis of residential assimilation. *International Migration Review.* Forthcoming.

Myers, D., and Wolch, J. 1994. The polarization of housing status. In *State of the Union,* R. Farley (Ed.), pp. 269–356. New York: Russell Sage Foundation.

NALEO Educational Fund. 1989. *The National Latino Immigrant Survey.* Washington, DC: National Association of Latino Elected Officials (NALEO) Educational Fund.

NALEO Educational Fund. 1992. *National Register of Hispanic Elected Officials.* Los Angeles: National Association of Latino Elected Officials Educational Fund.

National Asian Pacific American Political Almanac. 1996. Los Angeles, CA: UCLA Asian American Studies Center.

National Research Council. 1997. *The New Americans: Economic, Demographic, and Fiscal Effects of Immigration.* Washington, DC: National Academy Press.

Navarro, J. 1994. The economic effects of emigration: Mexico. In *Emigration and Its Effects on the Sending Country*, B. Asch (Ed.), pp. 185–204. Santa Monica, CA: Rand Corporation.

Neuman, K., and Tienda, M. 1994. The settlement and secondary migration patterns of legalized immigrants: Insights from administrative records. In *Immigration and Ethnicity: The Integration of America's Newest Arrivals*, B. Edmonston and J. Passel (Eds.), pp. 187–226. Washington, DC: Urban Institute Press.

Oliver, M., and Shapiro, T. 1995. *Black Wealth/White Wealth*. New York: Routledge.

Olsen, E. 1987. The demand and supply of housing. In *Handbook of Regional Housing Economics*, E. Mills (Ed.), Vol. 2, pp. 989–1022. Amsterdam: Elsevier Science Publishers.

Ong, P., Bonacich, E., and Chang, L. 1994. *The New Asian Immigration in Los Angeles and Global Restructuring*. Philadelphia: Temple University Press.

Ortiz, V. 1996. The Mexican-origin population: Permanent working class or emerging middle class? In *Ethnic Los Angeles*, R. Waldinger and M. Bozorgmehr (Eds.), pp. 247–278. New York: Russell Sage Foundation.

Papademetriou, D. G., and Yale-Loehr, S. 1996. *Balancing Interests: Rethinking U.S. Selection of Skilled Immigrants*. Washington, DC: Carnegie Endowment for International Peace.

Pasqualucci, J. M. 1994–1995. The inter-American human rights system: Establishing precedents and procedures in human rights law. *University of Miami Inter-American Law Review* 26: 363–401.

Passel, J., and Clark, R. 1994. *How Much Do Immigrants Really Cost? A Reappraisal of Huddle's "The Cost of Immigrants."* Washington, DC: Urban Institute Press.

Passell, J., and Woodrow, K. 1984. Geographic distribution of undocumented immigrants: Estimates of undocumented Mexicans counted in the U.S. Census by state. *International Migration Review* 18: 642–671.

Pinal, J., and Singer, A. 1997. *Generations of Diversity: Latinos in the United States*. Washington, DC: Population Reference Bureau.

Pincetl, S. 1994. Challenges to citizenship: Latino immigrants and political organizing in the Los Angeles area. *Environment and Planning A* 26: 895–914.

Plyler v. Doe. 1982. 457 U.S. 202, 210.

Population Reference Bureau. 1997. *World Population Data Sheet*. Washington, DC: Population Reference Bureau.

Porter, R. 1990. *Forked Tongue: The Politics of Bilingual Education*. New York: Basic Books.

Portes, A., and Rumbaut, R. 1996. *Immigrant America: A Portrait*. Berkeley and Los Angeles: University of California Press.

Portes, A., and Zhou, M. 1994. Should immigrants assimilate? *Public Interest* 116: 18–33.

Ravitch, D. 1990. Multiculturalism: E pluribus plures. *American Scholar* 59: 337–354.

Reed, D., Haber, M. G., and Mameesh, L. 1996. *The Distribution of Income in California*. San Francisco: Public Policy Institute of California.

Reitz, J. 1998. *Warmth of the Welcome: The Social Causes of Economic Success for Immigrants in Different Nations and Cities.* Boulder, CO: Westview Press.

Rieff, D. 1991. *Los Angeles: Capital of the Third World.* New York: Simon & Schuster.

Root, M. P. 1992. *Racially Mixed People in America.* Newbury Park, CA: Sage.

Rodriguez, G. 1996. *The Emerging Latino Middle Class.* Los Angeles: Pepperdine University.

Rolph, E. 1992. *Immigration Policies: Legacy from the 1980's and Issues for the 1990's.* Santa Monica, CA: Rand Corporation.

Rose, P. 1993. Of every hue and caste: race, immigration and perceptions of pluralism. *Annals of the Association of Political and Social Science* 530: 187–202.

Rose Institute. 1988. *California's Latinos, 1988: An Opinion Survey.* Claremont, CA: Claremont McKenna College.

Rossell, C., and Baker, K. 1996. The educational effectiveness of bilingual education. *Research in the Teaching of English* 30: 7–74.

Rotberg, I. 1982. Federal policy in bilingual education. *American Education* 52: 30–40.

Rumbaut, R. 1997. *Passages to Adulthood: The Adaptation of Children of Immigrants in Southern California.* New York: Russell Sage Foundation.

Saad, L. 1995, July. Immigrants see U.S. as land of opportunity. *Gallup Poll Monthly,* pp. 19–33.

Sassen, S. 1991. *The Global City.* Princeton, NJ: Princeton University Press.

Sassen, S. 1994. *Cities in a World Economy.* London: Pine Forge Press.

Sawhill, I. 1988. Poverty in the U.S.: Why is it so persistent? *Journal of Economic Literature* 26: 1073–1119.

Schlesinger, A., Jr. 1992. *The Disuniting of America: Reflections on a Multi-Cultural Society.* New York: W.W. Norton.

Schuck, P. 1990, Fall. The great immigration debate. *American Prospect,* pp. 100–118.

Shoeni, R., McCarthy, K., and Vernez, G. 1996. *The Mixed Economic Progress of Immigrants.* Santa Monica, CA: Rand Corporation.

Simon, J. 1984. Immigrants, taxes, and welfare in the United States. *Population and Development Review* 10: 55–70.

Simon, J. 1989. *The Economic Consequences of Immigration.* Cambridge, MA: Basil Blackwell.

Simon, J., and Akburi, A. 1995. Educational trends of immigrants in the U.S. Unpublished paper, meetings of the Population Association of America.

Sjaastad, L. 1961. The costs and returns of human migration. *Journal of Political Economy* 70: 80–93.

Skerry, P. 1993. *Mexican Americans: The Ambivalent Minority.* New York: The Free Press.

Smith, J., and Edmonston, B. 1997. *The New Americans: Economic, Demographic and Fiscal Effects of Immigration.* Washington, DC: National Academy of Sciences Press.

Sorenson, E., and Bean, F. 1994. The Immigration Reform and Control Act and the wages of Mexican-origin workers: Evidence from current population surveys. *Social Science Quarterly* 75 (1): 1–17.

Sorenson, E., and Enchautegui, M. 1994. Immigrant male earnings in the 1980s: Divergent patterns by race and ethnicity. In *Immigration and Ethnicity: The Integration of America's Newest Arrivals,* B. Edmonston and J. Passel (Eds.), pp. 139–162. Washington, DC: Urban Institute Press.

Spickard, P. R. 1989. *Mixed Blood: Intermarriage and Ethnic Identity in Twentieth-Century America.* Madison, WI: University of Wisconsin Press.

Stark, O., and Bloom, D. E. 1985. The new economics of labor migration. *American Economic Review* 75: 173–178.

Takaki, R. 1993. Multiculturalism: Battleground or meeting ground? *Annals of the Association of Political and Social Science* 530: 122–136.

Taylor, E., Martin, P., and Fix, M. 1997. *Poverty amid Prosperity: Immigration and the Changing Face of Rural California.* Washington, DC: Urban Institute Press.

Thernborn, G. 1995. *European Modernity and Beyond: The Trajectory of European Societies, 1945–2000.* Thousand Oaks, CA: Sage.

Thom, L. 1995. How Thor tried to drain the magic drinking horn, or why poverty increases in the United States. *Population and Environment: A Journal of Interdisciplinary Studies* 17: 7–17.

U.S. Bureau of the Census. 1972. *U.S. Census of Population and Housing, 1970. Public Use Samples of Basic Records (California).* Washington, DC: Bureau of the Census.

U.S. Bureau of the Census. 1983. *U.S. Census of Population and Housing, 1980. Public Use Microdata Samples (California).* Washington, DC: Bureau of the Census.

U.S. Bureau of the Census. 1991. *U.S. Census of Population and Housing, 1990. Summary Tape File 1 (California).* Washington, DC: Bureau of the Census.

U.S. Bureau of the Census. 1992. *U.S. Census of Population and Housing, 1990. Public Use Microdata Samples (California).* Washington, DC: Bureau of the Census.

U.S. Bureau of the Census. 1993. *U.S. Census of Population and Housing, 1990. Summary Tape File 3 (California).* Washington, DC: Bureau of the Census.

U.S. Commission on Immigration Reform. 1997. *Becoming an American: Immigration and Immigrant Policy.* Washington, DC: U.S. Commission on Immigration Reform.

U.S. Department of Commerce. 1996. *Statistical Assessment of the United States.* Washington, DC: Bureau of the Census.

U.S. Department of Education. 1993. *Prospects: The Congressionally Mandated Study of Educational Growth and Opportunity.* Washington, DC: U.S. Department of Education.

U.S. Department of Education. 1997. *Dropout Rates in the United States.* Washington, DC: National Center for Education Statistics.

U.S. Immigration and Naturalization Service. 1996. *Immigration to the United States in Fiscal Year 1995.* Washington, DC: U.S. Department of Justice.

Van Hook, J., and Bean, F. 1998. Estimating unauthorized migration to the United States: Issues and results. In *Mexico–United States Binational Migration Study.* Washington, DC: U.S. Commission on Immigration Reform.

Vaughan, J. 1997, Spring. Immigrant visa waiting list at 3.6 million. *Immigration Review* 28: 5–6.

Vernez, G., and Abrahamse, A. 1996. *How Immigrants Fare in U.S. Education.* Santa Monica, CA: Rand Corporation.

Waldinger, R. D. 1986. *Through the Eye of the Needle: Immigrants and Enterprise in New York's Garment Trades.* New York: New York University Press.

Waldinger, R. D. 1996. *Still the Promised City? African Americans and New Immigrants in Postindustrial New York.* Cambridge, MA: Harvard University Press.

Waldinger, R. D., and Bozorgmehr, M. (Eds.). 1996. *Ethnic Los Angeles.* New York: Russell Sage Foundation.

Waldorf, B. 1994. Assimilation and attachment in the context of international migration. *Papers in Regional Science* 73: 241–266.

Waldorf, B. 1996. The internal dynamic of international migrations systems. *Environment and Planning A* 28: 631–650.

Waldorf, B., Esparaza, A., and Huff, J. 1990. A behavioral model of international labor and nonlabor migration: The case of Turkish movements to West Germany, 1960–1986. *Environment and Planning A* 41: 172–183.

Warren, R. 1997, April. *Report on the Size of the Illegal Population.* Conference paper, Annual meeting of the Population Association of America, Washington, DC.

Warren, R., and Passel, J. 1987. A count of the uncountable: Estimates of undocumented aliens counted in the 1980 United States Census. *Demography* 24: 375–393.

Weisbrod, B. 1965. *The Economics of Poverty: An American Paradox.* Englewood Cliffs, NJ: Prentice-Hall.

Western, B., and Kelly, E. 1997. Comparing demographic and labor-market characteristics of New Jersey and U.S. foreign born. In *Keys to Successful Immigration,* T. J. Espenshade (Ed.), pp. 35–54. Washington, DC: Urban Institute Press.

Wilson, W. J. 1987. *The Truly Disadvantaged.* Chicago: University of Chicago Press.

Woodrow, K., and Passel, J. 1990. Post IRCA undocumented immigration to the United States: An assessment based on the 1988 CPS. In *Undocumented Migration to the United States: IRCA and the Experience of the 1980s,* F. D. Bean, B. Edmunston, and J. Passel (Eds.), pp. 33–76. Santa Monica, CA: Rand Corporation.

Woodrow-Lafield, K. 1994. *A Sociology of Official Statistics on Undocumented Immigrants.* Albany: State University of New York Press.

Wright, R., and Ellis, M. 1996. Immigrants and the changing racial/ethnic division of labor in New York City. *Urban Geography* 17: 317–353.

Wright, R., Ellis, M., and Reibel, M. 1997. The linkage between immigration and internal migration in large metropolitan areas in the United States. *Economic Geography* 73: 234–254.

Zill, N., and Collins, M. 1995. *Approaching Kindergarten: A Look at Preschoolers in the United States.* Washington, DC: National Center for Health Statistics.

INDEX